I0464601

Introduction to the Naval Research Laboratory

Mission

To conduct a broadly based multidisciplinary program of scientific research and advanced technological development directed toward maritime applications of new and improved materials, techniques, equipment, systems, and ocean, atmospheric, and space sciences and related technologies.

The Naval Research Laboratory

- Provides primary in-house research for the physical, engineering, space, and environmental sciences;

- Provides broadly based exploratory and advanced development programs in response to identified and anticipated DON needs;

- Provides broad multidisciplinary support to the Naval Warfare Centers;

- Provides space and space systems technology development and support; and

- Assumes responsibility as the Navy's corporate laboratory.

The Naval Research Laboratory is located in Washington, DC, on the east bank of the Potomac River.

The NRL Marine Meteorology Division is located in Monterey, California (NRL-MRY).

The Naval Research Laboratory Detachment is located at Stennis Space Center, Bay St. Louis, Mississippi (NRL-SSC).

1

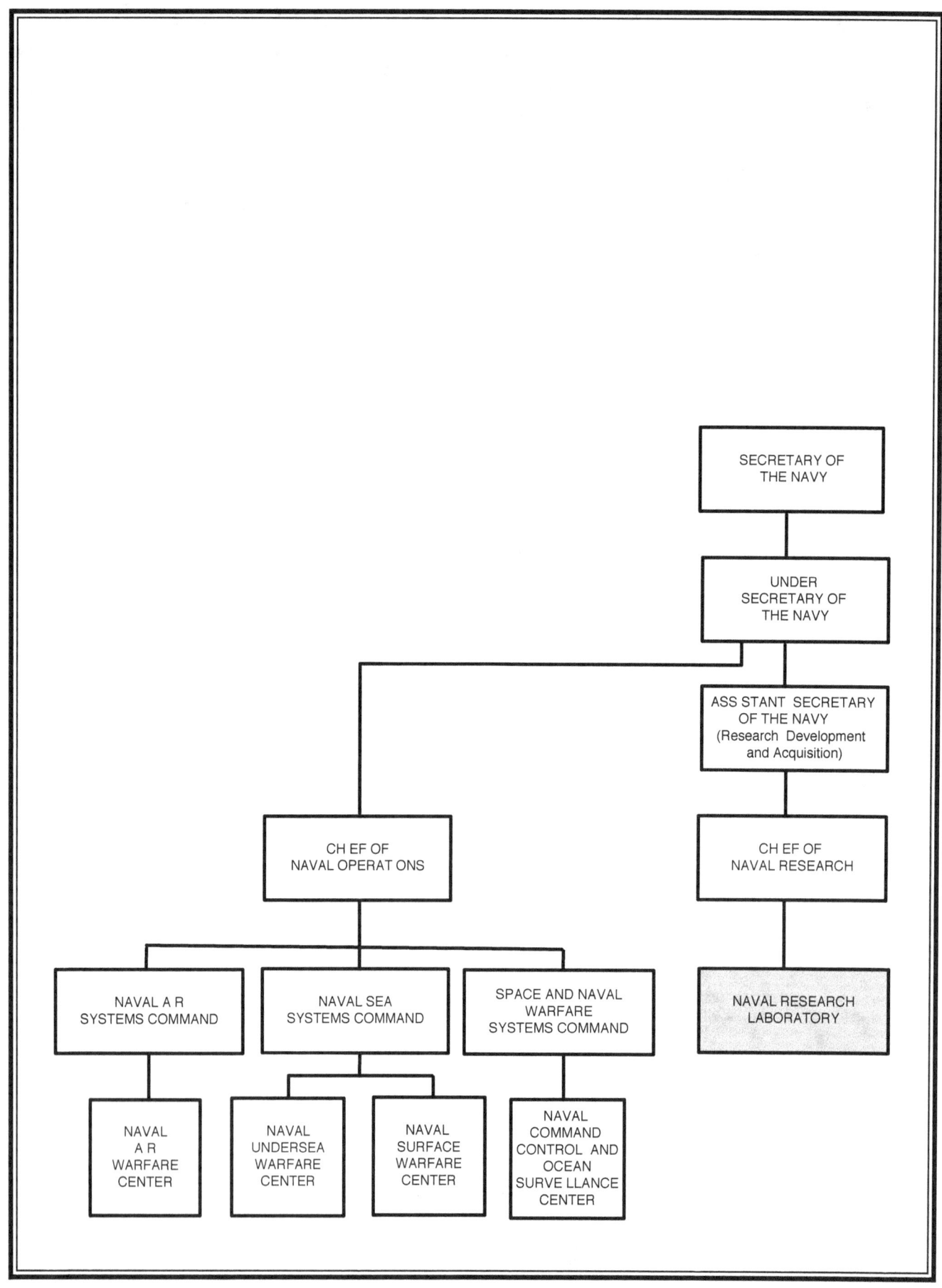

SECRETARY OF
THE NAVY

UNDER
SECRETARY OF
THE NAVY

ASS STANT SECRETARY
OF THE NAVY
(Research Development
and Acquisition)

CH EF OF
NAVAL OPERAT ONS

CH EF OF
NAVAL RESEARCH

NAVAL A R
SYSTEMS COMMAND

NAVAL SEA
SYSTEMS COMMAND

SPACE AND NAVAL
WARFARE
SYSTEMS COMMAND

NAVAL RESEARCH
LABORATORY

NAVAL
A R
WARFARE
CENTER

NAVAL
UNDERSEA
WARFARE
CENTER

NAVAL
SURFACE
WARFARE
CENTER

NAVAL
COMMAND
CONTROL AND
OCEAN
SURVE LLANCE
CENTER

The Naval Research Laboratory
in the
Department of the Navy

The Naval Research Laboratory is the Department of the Navy's corporate laboratory; it is under the command of the Chief of Naval Research. As the corporate laboratory of the Navy, NRL is the principal in-house component in the Office of Naval Research's (ONR) effort to meet its science and technology responsibilities.

NRL has had a long and fruitful relationship with industry as a collaborator, contractor, and most recently in Cooperative Research and Development Agreements (CRADAs). NRL values this linkage and continues to develop it.

NRL is an important link in the Navy Research, Development, and Acquisition (RD&A) chain. Through NRL, the Navy has direct ties with sources of fundamental ideas in industry and the academic community throughout the world and provides an effective coupling point to the R&D chain for ONR.

NRL Functional Organization

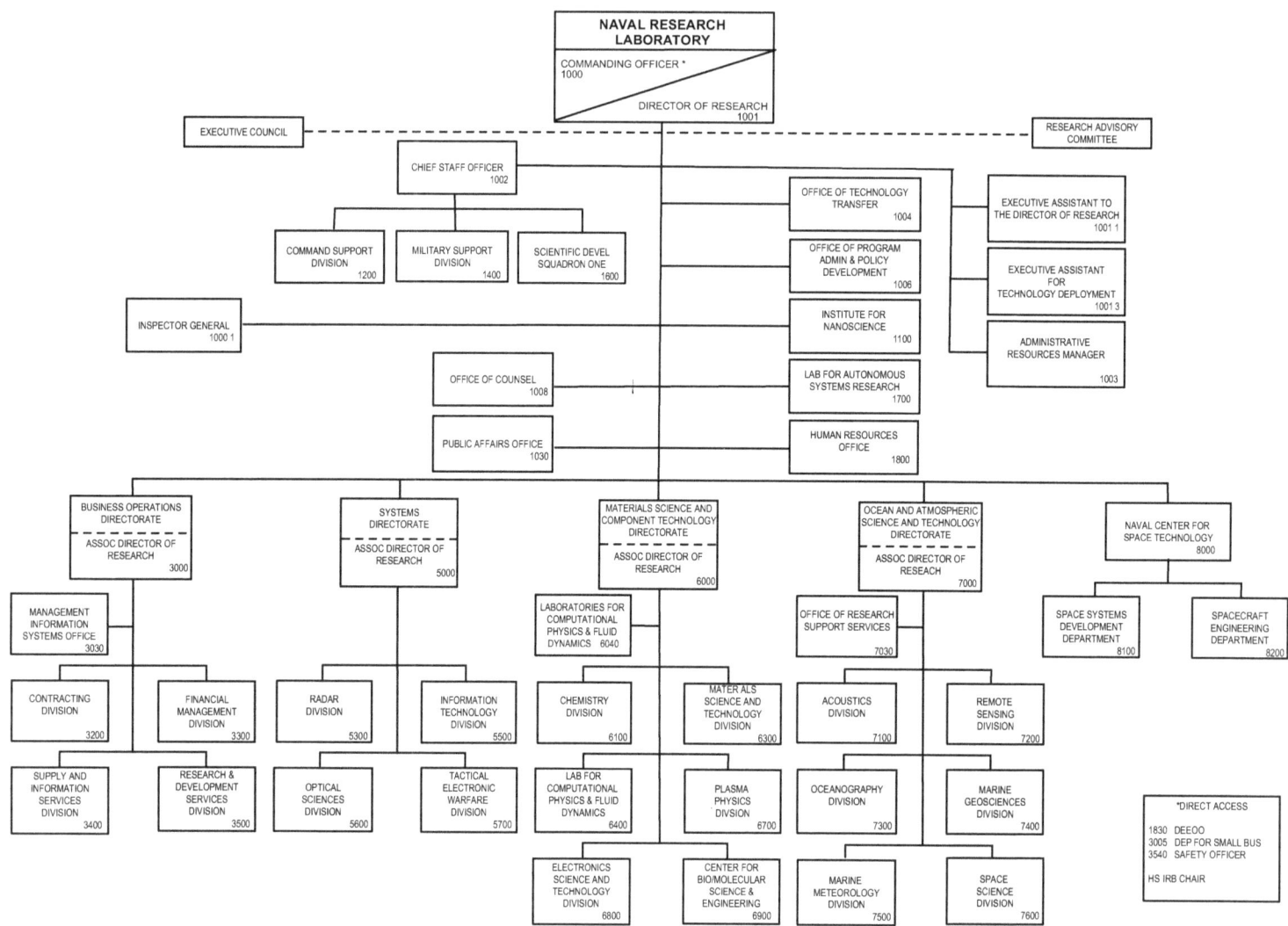

Current Research

The following areas represent broad fields of NRL research. Under each, more specific topics that are being investigated for the benefit of the Navy and other sponsoring organizations are listed. Some details of this work are given in the *NRL Review*, published annually. More specific details are published in reports on individual projects provided to sponsors and/or presented as papers for professional societies or their journals.

Advanced Radio, Optical, and IR Sensors

Advanced optical sensors
EM/EO/meteorological/oceanographic sensors
Satellite meteorology
Precise space tracking
Radio/infrared astronomy
Infrared sensors and phenomenology
UV sensors and middle atmosphere research
Image processing
VLBI/astrometry
Optical interferometry
Imaging spectrometry
Liquid crystal technology

Autonomous Systems

Algorithms for control of autonomous systems
Cognitive robotics
Human-robot interaction
Perception hardware and algorithms
High-level reasoning algorithms
Machine learning and adaptive algorithms
Sensors for autonomous systems
Power and energy for autonomous systems
Networking and communications for mobile systems
Swarm behaviors
Test and evaluation of autonomous systems

Computer Science and Artificial Intelligence

Standard computer hardware, development environments, operating systems, and run-time support software
Methods of specifying, developing, documenting, and maintaining software
Human-computer interaction
Intelligent systems for resource allocation, signal identification, operational planning, target classification, and robotics
Parallel scientific libraries
Algorithms for massively parallel systems
Digital progressive HDTV for scientific visualization
Adaptive systems: software and devices
Advanced computer networking
Simulation management software for networked high performance computers
Interactive 3D visualization tools and applications
Real-time parallel processing
Scalable, parallel computing
Petaflop computing, globally distributed file systems, terabit-per-second networking

Directed Energy Technology

High-energy lasers
Laser propagation
Solid-state and fiber lasers
High-power microwave sources
RAM accelerators
Pulse detonation engines
Charged-particle devices
Pulse power
DE effects

Electronic Electro-optical Device Technology

Integrated optics
Radiation-hardened electronics
Nanotechnology
Microelectronics
Microwave and millimeter-wave technology
Hydrogen masers for GPS
Aperture syntheses
Electric field coupling
Vacuum electronics
Focal plane arrays
Infrared sensors
Radiation effects and satellite survivability
Molecular engineering

Electronic Warfare

EW/C2W/IW systems and technology
COMINT/SIGINT technology
EW decision aids and planning/control systems
Intercept receivers, signal processing, and identification systems
Passive direction finders
Decoys and offboard countermeasures (RF and IR)
Expendable autonomous vehicles/UAVs
Repeaters/jammers and EO/IR active countermeasures and techniques
Platform signature measurement and management
Threat and EW systems computer modeling and simulations
Visualization
Hardware-in-the-loop and flyable ASM simulators
Missile warning infrared countermeasures
RF environment simulators
EO/IR multispectral/hyperspectral surveillance

Enhanced Maintainability, Reliability, and Survivability Technology

Coatings
Friction/wear reduction
Water additives and cleaners

Fire safety
Laser hardening
Satellite survivability
Corrosion control
Automation for reduced manning
Radiation effects
Mobility fuels
Chemical and biological sensors
Environmental compliance

Environmental Effects on Naval Systems

Meteorological effects on communications
Meteorological effects on weapons, sensors, and
platform performance
Air quality in confined spaces
Electromagnetic background in space
Solar and geomagnetic activity
Magnetospheric and space plasma effects
Nonlinear science
Ionospheric behavior
Oceanographic effects on weapons, sensors, and
platforms
EM, EO, and acoustic system performance/
optimization
Environmental hazard assessment
Contaminant transport
Biosensors
Microbially induced corrosion

Imaging Research/Systems

Remotely sensed signatures analysis
Real-time signal and image processing algorithms/
systems
Image data compression methodology
Image fusion
Automatic target recognition
Scene/sensor noise characterization
Image enhancement/noise reduction
Scene classification techniques
Radar and laser imaging systems studies
Coherent/incoherent imaging sensor exploitation
Remote sensing simulation
Hyperspectral imaging
Microwave polarimetry

Information Technology

High-performance, all-optical networking
Antijam communication links
Next-generation, signaled optical network
architectures
Integrated voice and data
Information security (INFOSEC)
Voice processing
High performance computing
High performance communications
Requirements specification and analysis
Real-time computing
Wireless mobile networking
Behavior detection
Machine learning

Information filtering and fusion
Integrated internet protocol (IP) and asynchronous
transfer mode (ATM) multicasting
Reliable multicasting
Wireless networking with directional antennas
Sensor networking
Communication network simulation
Bandwidth management (quality of service)
High assurance software
Distributed network-based battle management
High performance computing supporting uniform
and nonuniform memory access with single and
multithreaded architectures
Distributed, secure, and mobile information
infrastructures
Simulation-based virtual reality
High-end, progressive HDTV imagery processing
and distribution
Defensive information warfare
Virtual reality/mobile augmented reality
3D multimodal interaction
Model integration (physical, environmental,
biological, psychological) for simulation
Command decision support
Data fusion

Marine Geosciences

Marine seismology, including propagation and
noise measurement
Geoacoustic modeling in support of acoustic
performance prediction
Geomagnetic modeling in support of nonacoustic
system performance prediction
Static potential field measurement and analysis
(gravity and magnetic) in support of navigation
and geodesy
Geotechnology/sediment dynamics affecting mine
warfare and mine countermeasures
Foreshore sediment transport
Geospatial information, including advanced
seafloor mapping, imaging systems, and
innovative object-oriented digital mapping
models, techniques, and databases

Materials

Superconductivity
Magnetism
Biological materials
Materials processing
Advanced alloy systems
Solid free-form fabrication
Environmental effects
Energetic materials/explosives
Aerogels and underdense materials
Nanoscale materials
Nondestructive evaluation
Ceramics and composite materials
Thin film synthesis and processing
Electronic and piezoelectric ceramics
Thermoelectric materials

Active materials and smart structures
Computational material science
Paints and coatings
Flammability
Chemical/biological materials
Spintronic materials and half metals
Biomimetic materials
Multifunctional materials
Power and energy
Synthetic biology

Meteorology
Global, theater, tactical-scale, and on-scene
    numerical weather prediction
Data assimilation and physical initialization
Atmospheric predictability and adaptive
    observations
Adjoint applications
Marine boundary layer characterization
Air/sea interaction; process studies
Coupled air/ocean/land model development
Tropical cyclone forecasting aids
Satellite data interpretation and application
Aerosol transport modeling
Meteorological applications of artificial
    intelligence and expert systems
On-scene environmental support system
    development/nowcasting
Tactical database development and
    applications
Meteorological tactical decision aids
Meteorological simulation and visualization

Ocean Acoustics
Underwater acoustics, including propagation,
    noise, and reverberation
Fiber-optic acoustic sensor development
Deep ocean and shallow water environmental
    acoustic characterization
Undersea warfare system performance
    modeling, unifying the environment,
    acoustics, and signal processing
Target reflection, diffraction, and scattering
Acoustic simulations
Tactical decision aids
Sonar transducers
Dynamic ocean acoustic modeling

Oceanography
Oceanographic instrumentation
Open ocean, littoral, polar, and nearshore
    oceanographic forecasting
Shallow water oceanographic effects on
    operations
Modeling, sensors, and data fusion
Bio-optical and fine-scale physical processes
Oceanographic simulation and visualization
Coastal scene generation
Waves, tides, and surf prediction
Coupled model development

Coastal ocean characterization
Oceanographic decision aids
Global, theater, and tactical-scale modeling
Remote sensing of oceanographic parameters
Satellite image analysis

Space Systems and Technology
Space systems architectures and requirements
Advanced payloads and optical communications
Controllers, processors, signal processing, and VLSI
Precision orbit estimation
Onboard autonomous navigation
Satellite ground station engineering and
    implementation
Tactical communication systems
Spacecraft antenna systems
Launch and on-orbit support
Precise Time and Time Interval (PTTI) technology
Atomic time/frequency standards/instrumentation
Passive and active ranging techniques
Design, fabrication, and testing of spacecraft and
    hardware
Structural and thermal analysis
Attitude determination and control systems
Reaction control
Propulsion systems
Navigation, tracking, and orbit dynamics
Spaceborne robotics applications

Surveillance and Sensor Technology
Point defense technology
Imaging radars
Surveillance radars
Multifunction RF systems
High-power millimeter-wave radar
Target classification/identification
Airborne geophysical studies
Fiber-optic sensor technology
Undersea target detection/classification
EO/IR multispectral/hyperspectral detection and
    classification
Sonar transducers
Electromagnetic sensors, gamma ray to RF
    wavelengths
SQUID for magnetic field detection
Low observables technology
Ultrawideband technology
Interferometric imagery
Microsensor system
Digital framing reconnaissance canvas
Biologically based sensors
Digital radars and processors

Undersea Technology
Autonomous vehicles
Bathymetric technology
Anechoic coatings
Acoustic holography
Unmanned undersea vehicle dynamics
Weapons launch

Major Research Capabilities and Facilities
(Listed alphabetically by organizational unit)

Acoustics Division (Code 7100)
Laboratory Measurements
- One-million-gallon, vibration-isolated underwater acoustic holographic/3D laser vibrometer facility for studying structural acoustic phenomena
- Large, sandy-bottom, acoustic holographic pool facility for investigating echo characteristics of underwater buried/near-bottom targets and sediment acoustics
- In-air structural acoustics facility with high spatial density near-field acoustic holography and 3D laser vibrometry for diagnosing large structures, including aircraft interiors and rocket payload fairings
- Salt water acoustic tank (20 ft by 20 ft by 10 ft deep) with environmental control and substantial optical access for studying the acoustics of bubbly media, acoustic metamaterials, and laser induced sound
- Micro-Nanostructure Dynamics Laboratory to study the structural dynamics and performance of high Q oscillators and other micromechanical systems using laser Doppler vibrometers, super resolution nearfield scanning optical microscope, and low temperature calorimeter
- Model Fabrication Laboratory to fabricate rough topographical surfaces in various materials for acoustic scattering and propagation studies and measurements.
- Sonomagnetic Laboratory with doubly insulated Faraday cage for conducting experiments to measure weak electromagnetic fields generated by mechanical/acoustic vibrations of a conducting medium in an arbitrary magnetic field

Seagoing Assets
- Acoustic arrays (towed/moored/suspended)
- 64-channel broadband source–receiver array with time-reversal mirror functionality over a frequency band of 500 to 3500 Hz
- High-powered sound sources and source arrays
- Autonomous acoustic sources
- Acoustic communications array and data acquisition buoy
- Portable, ocean-deployable synthetic aperture acoustic measurement system (100-meter rail with precise positioning)
- Containerized, seagoing multichannel data acquisition system
- High-speed, maneuverable towed body with MK-50 and synthetic aperture sonars to measure high frequency scattering and coherence

Center for Bio/Molecular Science and Engineering (Code 6900)
Optical equipment
- Confocal microscope
- Raman microscope
- UV-visible absorption spectrophotometers
- Transmission electron microscope
- Scanning electron microscope
- Microscope/atomic force microscope
- Nanosight (nanoparticle tracking analysis)

Analytical instruments
- Gas chromatography mass spectrometer
- HPLC
- LC/MS/MS system
- FluroMax-3 spectrofluorometer
- Titration workstation

General facilities
- X-ray scattering
- Cold room for storage and preparation
- High-speed ultracentrifuges
- Inert atmosphere dry box
- NMR
- FTIR
- Ellipsometer
- Dynamic mechanical analyzer
- Differential scanning calorimeter
- Circular dichroism
- Minimill injection mold machine
- Multi RF centrifuge
- Perkin Elmer BioChip Arrayer I
- Freeze-dry system
- Affymetrix Gene Chip system
- Surface plasmon resonance (SPR)
- Isothermal calorimeter

Chemistry Division (Code 6100)
Synthesis/processing facilities
- Paint formulation and coating
- Functional polymers/elastomers/composites
- Nanotubes/Nanofibers
- Surface modification
- Thin film deposition/etching with in situ control

Marine Corrosion Facility (at Key West, FL)
Fire/Damage Control Test Facility (at Mobile, AL)
Characterization facilities
- General-purpose chemical analysis/trace analysis
- Surface diagnostics
- Nanometer scale composition/structure/properties
- Magnetic resonance NDI
- Tribology
- Polymer structure/function/dynamics

Special-purpose capability
- Environmental monitoring/remediation
- Combustion and fire research
- Alternate and petroleum-derived fuels
- Trace explosive detection test beds
- Trace vapor generation and detection test beds

Simulation/modeling
Synchrotron radiation beam lines (at NSLS, Brookhaven, NY)
Pressurized test chambers (small, medium, large)

Electronics Science and Technology Division (Code 6800)

Nano- and microelectronics characterization and processing facilities

Electron-beam nanowriter

High-resolution transmission electron microscope

Scanning tunneling microscopy and electro-optical analysis

Material growth facilities including bulk crystal growth, molecular beam epitaxy, organometallic chemical vapor deposition, and atomic layer deposition

Optical and electrical characterization of materials

Electronic testing and analysis facilities

Cathode fabrication and characterization laboratory

Millimeter-wave vacuum electronics fabrication facility

Femtosecond laser facility

Solar cell characterization facility

Power electronics materials characterization and device processing facilities

Information Technology Division (Code 5500)

Extended Spectrum Experimentation Laboratory

Robotics and Autonomous Systems Laboratory

Immersive Simulation Laboratory

Warfighter Human-Systems Integration Laboratory

Audio Laboratory

Mobile and Dynamic Network Laboratory

Integrated Communications Technology Test Lab

General Electronics Environmental Test Facility

Key Management Laboratory

Crypto Technology Laboratory

Navy Cyber Defense Research Laboratory

Communications Security (COMSEC) Laboratory

Navy Shipboard Communications Testbed

Behavior Detection Laboratory

Virtual Reality Laboratory

Service Oriented Architecture Laboratory

Distributed Simulation Laboratory

Motion Imagery Laboratory

Laboratory for Large Data Research

Affiliated Resource Center for High Performance Computing

Ruth H. Hooker Research Library

Institute for Nanoscience (Code 1100)

Clean room (5000 sq ft), quiet (4000 sq ft), and ultra-quiet (1000 sq ft) laboratories

35 dB and 25 dB acoustically isolated zones

$20°C \pm 0.5°C$ and $0.1°C$ controlled temperature zones

Vibration isolation

Vertical (mm, pp) <0.1 @ 70–500 Hz

Horizontal (mm, pp) <0.1 @ 70–500 Hz

Clean electrical power, free from SCR spikes and other interferences, and $< \pm 10\%$ voltage change

<0.5 mG at 60 Hz EMI

$45 \pm 5\%$ relative humidity

Class 100 clean room

Source of water meeting ASTM D5127 spec. Type E1.2

Clean Room Major Equipment

Monitoring system (toxic gas, hazmat, temperature)

Laminar flow wet benches for localized Class 1/10 ambient in clean room

Air purification unit to remove local organic contamination

DI water system

Wire bonder

E-beam writer with active vibration control system

Scanning electron microscope

Atomic force microscope

Metallurgical optical microscopes

Surface profiler

Mask aligners (2, 1, and 0.2 µm)

Electron beam evaporation system

Low pressure chemical vapor deposition (LPCVD) system

Magnetron sputter deposition system

Reactive ion etching systems

Dual-beam focused ion beam workstation

Optical pattern generating system

Laser micromachining system

Plasma-enhanced chemical vapor deposition (PECVD) system

Plasma-enhanced atomic layer deposition system

Chlorine reactive ion etching system

Other Major Equipment

Transmission electron microscope

UHV multi-tip scanning tunneling microscope/ nanomanipulator

Laboratories for Computational Physics and Fluid Dynamics (Code 6040)

1120-core x86 cluster

(3) 64-core SGI Altix systems

184-core x86 cluster

256-core SGI ICE

256-processor Opteron cluster

More than sixty SGI, Apple, and Intel workstations

Three-quarter-terabyte RAID disk storage systems

All computers and workstations have network connections to NICENET and ATDnet allowing access to the NRL CCS facilities (including the DoD HPC resources) and many other computer resources both internal and external to NRL

Laboratory for Autonomous Systems Research (Code 1700)

Prototyping High Bay: (150 ft by 75 ft by 30 ft), contains real-time motion capture system, directional environmental sounds, GPS repeater and simulator

Four human-systems interaction labs contain eye trackers and multiuser, multitouch monitors

Littoral High Bay with 45 ft by 25 ft by 5.5 ft deep pool with 16-channel wave generator and slope that allows simulation of littoral environments; multiple sediment tanks (from 5 ft to 16 ft); GPS repeater and simulator; portable tank 4 ft by 36 ft

Desert High Bay with a 40 ft by 14 ft area of sand 2.5 ft deep, and 18 ft high rock walls; high speed fans and variable lighting

Tropical High Bay, a 60 ft by 40 ft greenhouse, contains a re-creation of a southeast Asian rain forest with native plants; nominal 80 degrees temperature and 80% humidity; can generate rain events up to 6 in. per hour; Rainforest contains waterfall, stream, and pond

Outdoor test range is a 1/3 acre highland forest with a waterfall, stream and pond, and terrain of differing difficulty including large bolder structures and earthen berms

Sensor lab contains environmental chambers (small and walk-in) with maximum temperature range of −50°F to 375°F, relative humidity from 10% to 95% and for smaller chamber, barometric pressure of −9000 feet to 100,000 feet; lab also contains various fume hoods, biosafety cabinet, anechoic chamber, vapor generators, and other specialized equipment

Power and energy lab contains specialized equipment including a battery dry room, glove box, isolation room, and fume hoods

Marine Geosciences Division (Code 7400)

Airborne gravimetry, magnetics, and topographic measurements suite coupled with differential GPS yielding position accuracies of <1.0 meter

100 and 500 kHz sidescan sonar with 2–12 kHz chirp profiler and Cs magnetometer for seafloor characterization/imaging and shallow subbottom profiling

Deep-towed acoustic geophysical system operating at 220–1000 Hz characterizes subseafloor structure including gas clathrate accumulations and dissociation of methane hydrates

Acoustic seafloor classification system operating at 8–50 kHz provides underway, real-time prediction of sediment type and physical properties

Seafloor probes for measuring sediment pore water pressures, permeability, electrical resistivity, acoustic compressional and shear wave velocities and attenuations, and dynamic penetration resistance

100 and 300 kV transmission electron microscopes with environmental cell for study of sediment fabric, especially impact of organic matter

Map data formatting facility compresses map information onto CD-ROM media for masters for use in aircraft digital moving map systems

Comprehensive geotechnical and geoacoustics laboratory capability

Airborne electromagnetic (AEM) bathymetry system

Ocean bottom magnetometer system

3D, multispectral, subbottom swath imaging system

Ocean bottom seismographs (OBS)

In situ sediment acoustic measurement system (IS-SAMS)

Instrumented mine shapes to measure hydrodynamics of free-fall in the water column, dynamics of deceleration in seafloor sediments, and rates and depths of scour burial

Hydrothermal plume imaging data acquisition and analysis system

Integrated digital databases analysis and display system for bathymetric, meteorological, oceanographic, geoacoustic, and acoustic data

Stereometric video image processing system for use in foreshore morphology measurement

Sediment gas-content sampler

Acoustic tomographic probes for surf zone sands and gassy muds

Computed tomography (CT) system and real-time radiography unit with a 0–225 keV @ 0–1 mA microfocus X-ray tube and a 225 mm image intensifier

Patented Geospatial Information Data Base (GIDB™) for rapidly accessing disparate geospatial content on the Internet. This is the most extensive interconnection of geospatial data that exists. http://dmap.nrlssc.navy.mil

Human-centered display design through the application of human factors principles in the design of geospatial displays (e.g., analysis of clutter in electronic displays)

GPS-based survey vehicles and equipment to measure foreshore and nearshore bathymetry (camera towers, jet ski, and push cart)

Geospatial visualization lab for rapid 2D and 3D graphic and physical visualization, analysis, and prototyping

Small oscillatory flow tunnel to observe sediment dynamics under forcing from waves and currents

Tomographic particle image velocimetry system for three-dimensional volumetric velocity measurements of fluid flow

Marine Meteorology Division (Code 7500)

The USGODAE Data Server (Global Ocean Data Assimilation Experiment) for collection and distribution of near-real-time METOC data and higher-level products from Navy and other providers to the global ocean and atmospheric research community

AN/SMQ-11 shipboard antenna system for retrieving polar-orbiting satellite data

Geostationary satellite data direct readout and polar-orbiting satellite data processing center

Supercomputer for numerical weather prediction systems development

Master Environmental Library (MEL) implemented on superworkstations for archiving and distributing real-time and historical atmosphere/ocean databases

Bergen Data Center for extensive file serving on disks and research data backup/archival capability on tapes

Data visualization center for developing shipboard briefing tools, displaying observations and model output, and integrating meteorological parameters into tactical simulations

Classified radar and satellite data processing facility

Two Mobile Atmospheric Aerosol and Radiation Characterization Observatories (MAARCO)

Technical research library

Materials Science and Technology Division (Code 6300)

Hot isostatic press

Cold isostatic press

High-energy dispersive X-ray analytical system

Electron microprobe, SEM, SAM, and STEM systems

Quantitative metallography

Computer-controlled multiaxial loading and SCC measurement systems

Computer-aided experimental stress analysis

Crystallite orientation distribution function (CODF)

Class 1000 clean room; processing metallic film

Elevated temperature and structural characterization laboratory

Metallic film deposition systems

Magnetometry

Cryogenic facilities

High-field magnets

High-resolution analytical electron microscope

Isothermal heat treating facility

Vacuum arc melting facility

Vacuum induction melting facility

3 MeV tandem Van de Graaff accelerator

200 keV ion-implantation facility

Precision colorimeters

Polymer synthesis and characterization

Microwave device test facility

Excimer laser film deposition facility

Bomen infrared spectrometer facility

Diffuse light scattering facility

Femtosecond laser facility

Surface characterization facility

Accelerator mass spectrometry facility

Carbon-14 dating facility

Laminated object manufacturing system

Thermal analysis characterization suite (TGA/DSC/DMA/DEA/rheometer)

Dielectric characterization facility

Composites processing autoclave

3D ESPI strain measurement system

Biomechanical surrogate fabrication facility

Oceanography Division (Code 7300)

Towed sensor and advanced microstructure profiler systems for studying upper ocean fine and microstructure

Integrated absorption cavity and optical profiler systems for studying ocean optical characteristics

Self-contained bottom-mounted upward-looking acoustic profilers for measuring ocean variability

Acoustic Doppler profiler for determining ocean currents while under way

Remotely operated underwater vehicle (ROV)

Bottom-mounted acoustic Doppler profilers

Towed hyperspectral optical array

SCI processing facility

Satellite receiving stations for AVHRR, MODIS, and DMSP ocean color processing facility

Environmental scanning electron microscope, confocal laser scanning microscope, and the new Inspect S low vacuum scanning electron microscope for detailed studies of biocorrosion in naval materials

Real-time Ocean Observations and Forecast Facility for monitoring and tracking of ocean physical and bio-optical conditions

Slocum Electric Gliders for performing wide-area ocean surveys of temperature, salinity, and optical characteristics

SCANFISH MKII, a towed undulating vehicle system, designed for collecting 3D TS profile data of the water column

Bottom-mounted Shallow water Environmental Profiler in Trawl-safe Real-time configuration (SEPTR) for measuring temperature, salinity, and optical parameters in addition to current profiles and pressure

Optical Sciences Division (Code 5600)

Optical probes laboratory to study viscoelastic, structural, and transport properties of molecular systems

Short-pulse excitation apparatus for kinetic mechanisms investigations

IR laser facility for optical characterization of semiconductors

Facilities for synthesis and characterization of optical glass compositions and for the fabrication of optical fibers

Silica and IR fluoride/chalcogenide fiber fabrication facilities

Environmental testing of fiber sensors (acoustic, magnetic, electric field, etc.)

Laser diode pumped solid-state lasers

Mid-IR, low-phonon crystal growth facility

Infrared countermeasure techniques laboratory

Mobile, high-precision optical tracker

EO/IR technology/systems modeling and simulation capabilities

Field-qualified EO/IR measurement devices

Focal plane array evaluation facility

Facilities for fabricating and testing integrated optical devices

Panchromatic and multi- and hyperspectral digital imaging processing facilities

NRL P-3 aircraft sensor pallet

Airborne EO/IR and radar sensors

    VNIR through SWIR hyperspectral systems

    VNIR, MWIR, and LWIR high-resolution systems

    Wideband SAR systems

RF and laser data links

High-speed, high-power photodetector characterization

Communication link characterization to >100 Gbps

RF phase noise, noise figure, and network analysis

Ultrahigh-speed A/O converters

Plasma Physics Division (Code 6700)

Mercury, 6 MV, 360 kA, magnetically insulated inductive voltage adder

Gamble II, 1 MV, 1 MA pulsed power generator

HAWK, 1 MA inductive storage facility

Table-Top Terawatt (T³) laser system

Table-Top Ti: Sapphire Femtosecond Laser (TFL) systems (10 Hz and 1 kHz)

NIKE krypton fluoride laser facility

Space Physics Simulation Chamber

Plasma Applications Laboratory

Microwave facility for processing of advanced materials (2.45, 35, 83, and 60–120 GHz)

ELECTRA, test bed for high-rep 5 Hz KrF laser

Railgun Materials Testing Facility

Directed Energy Physics Facility

SWOrRD laser facility

Radar Division (Code 5300)

Shipboard radar research and development test beds:
  AMRFC test bed
  AN/SPS-49A(V)1

Airborne research radar facility, APS-137D(V)5

High-power 94 GHz radar system

Ultrahigh-resolution radar system (microwave microscope)

Radar signature calculation facility

Electromagnetic numerical computation facility

Compact range antenna measurement laboratory and nearfield scanner

Electronic protection (EP) and adaptive pulse compression (APC) test bed

Electronic computer-aided design facility

Microwave and RF instrumentation laboratory

Functional materials electromagnetic analysis laboratory

High-bandwidth, high-capacity data recording system

High frequency (HF) multiple-input-multiple-output (MIMO) test bed

Remote Sensing Division (Code 7200)

Ground-based water vapor millimeter-wave spectrometer (WVMS)

SAR processing facility

SCI processing facility
  SEALAB
  SAIL

Hyperspectral imaging, sensors, and processing

Optical remote sensing calibration lab/facility

Navy Prototype Optical Interferometer (NPOI)

NRL/NRAO 74 MHz Very Large Array

Free surface hydrodynamics laboratory (including a 10 m wave tank)

WindSat processing facility

Volume imaging lidar system

Aerosol and field measurement facility

NRL RP-3A aircraft sensors

Airborne polarimetric microwave imaging radiometer (APMIR)

Airborne lidar

Millimeter-wave imager

Interferometric synthetic aperture radar (InSAR)

Flight-level meteorological sensors

Visible/near infrared (VNIR) hyperspectral imaging systems

Mid-wave infrared (MWIR) indium antimonide (InSb) hyperspectral imaging system

Long-wave infrared (LWIR) quantum well IR photodetector (QWIP) imaging system

Research and Development Services Division (Code 3500)

Military construction

Research support engineering

Planning

Full range of facility contracting, including construction, architect/engineering services, facilities support, and reserved parking

Transportation

Telephone services

Maintenance and repair of buildings, grounds, and communication and alarm systems

Shops for machining, sheet metal, welding, and plating

Occupational safety and health

Environmental

Health physics

Spacecraft Engineering Department (Code 8200)

Chambers:
  Thermal-vacuum
  Acoustic reverberation
  Large, tapered horn, RF anechoic chamber
  EMI/EMC testing chamber

Facilities:
  Spacecraft high-reliability electronic and electrical rework facility
  Spacecraft electronic systems integration and test facility
  Radio frequency (RF) system development facility
  RF microcircuit fabrication clean room facility
  Large tapered horn RF anechoic chamber facility
  Frequency sources laboratory
  Shock and vibration test
  Clean rooms (multiple classes and sizes)
  Spacecraft fabrication and assembly
  Fuels testing
  Autoclave
  Space robotics laboratory
  Proximity operations testbed
  CAD/CAM
  Propulsion system welding

Static loads test
Star tracker characterization
Spacecraft spin balance
Modal analysis
Computational astrodynamic simulation and
visualization

Space Science Division (Code 7600)

Development and test facilities for satellite, sounding
rocket, and balloon instruments, to perform solar
terrestrial, astrophysical, astronomical, solar, upper/
middle atmospheric, and space environment sensing
Infrared Test Facility (IRTF)
Solar Coronagraph Optical Test Chamber (SCOTCH)
Vacuum Ultraviolet Calibration Facility (VUCF)
Gamma Ray Imaging Laboratory (GRIL)
Doppler Asymmetric Spatial Heterodyne Spectroscopy
(DASH) balloon instrument
Very high angular Resolution Imaging Spectrometer
(VERIS) sounding rocket instrument
Remote Atmospheric and Ionospheric Detection System
(RAIDS) International Space Station instrument
Extreme Ultraviolet Imaging Spectrometer (EIS) satel-
lite instrument
Sun Earth Connection Coronal and Heliospheric Inves-
tigation (SECCHI) satellite instrument suite
Solar Orbiter Heliospheric Imager (SoloHI) satellite
instrument
Wide-field Imager (WISPR) satellite instrument
Compact Coronograph (CCOR) satellite instrument
Special Sensor Ultraviolet Limb Imager (SSULI) satellite
instrument
Spatial Heterodyne Imager for Mesospheric Radicals
(SHIMMER) satellite instrument
Atmospheric Neutral Density Experiment (ANDE)
microsatellite
Extensive computer-assisted data manipulation, inter-
pretive, and theoretical capabilities for space sci-
ence instrumentation operations, data imaging, and
modeling
SECCHI Payload Operations Center (POC)
Fermi Gamma-ray Space Telescope (formerly GLAST)
Science Analysis Center (SAC)
Simulation of radiation detection and systems in space
and terrestrial environments (SWORD & SMART)
Mountain Wave Forecast Model (MWFM)
Advanced Level Physics High Altitude extension of the
Navy Operational Global Atmospheric Prediction
System (NOGAPS-ALPHA)
Synthetic Scene Generation Model (SSGM)
Integrating the Sun-Earth System for the Operational
Environment (ISES-OE)

Space Systems Development Department (Code 8100)

Payload test facility and processor development
laboratory
Laser communications and electro-optics
laboratories

Tactical Technology Development Laboratory
(TTDL)
Precision oscillator (clock) test facility
RF payload development laboratory with anechoic
chamber
Precision high-frequency RF compact range
anechoic chamber facility
Transportable ground station development,
assembly, and test facility
Multiplatform FPGA/ASIC/VLSI development
laboratory
Satellite telemetry, tracking, and satellite control at
Blossom Point, MD
L/C/S/X-band fixed antenna resources
Connectivity to the Air Force Satellite Control
Network (AFSCN)
Pomonkey field site: large antenna, space commu-
nications, and research facility
Midway Research Center space communications
and research facility
Optical telescope facility

Tactical Electronic Warfare Division (Code 5700)

Visualization display room
Transportable step frequency radar
Vehicle development laboratory
Offboard test platform
Compact antenna range facility
Isolation measurement chamber facility
RFCM techniques development chamber facility
Low-power anechoic chamber
High-power microwave research facility
Electro-optics mobile laboratory
Infrared-electro-optical calibration and character-
ization laboratory
Infrared missile simulator and simulator develop-
ment laboratory
Secure supercomputing facility
CBD/Tilghman Island IR field evaluation facility
Ultrashort pulse laser effects research and analysis
laboratory
Central Target Simulator facility
Flying Electronic Warfare laboratory
High-power RF explosive laboratory
Classified material lay-up facility
Classified computing facilities
RF measurement laboratory
Wet chemistry laboratory
Ultra-near-field test facility
RF and millimeter-wave laboratory
Optical laboratory
Paint room
Secure laboratories for classified projects

NRL Sites and Facilities

SITE	ACREAGE		BUILDINGS/STRUCTURES
	LAND OWNED/LEASED	EASEMENT/LICENSE-PERMIT	
District of Columbia			
NRL and Joint Base Anacostia-Bolling*	131/0	0/10.13	93/30
Virginia			
Midway Research Center Quantico*	162/0	0/0	7/11
Maryland			
NRL Scientific Development Squadron One (VXS-1), NAS Patuxent River*	Tenant		
Chesapeake Bay Section and Dock Facility Chesapeake Beach*	168/0	.6/.02	47/77
Multiple Research Site Tilghman Island*	3/0	0/0	3/3
Free Space Antenna Range Pomonkey*	141/0	0/0	10/10
Blossom Point Satellite Tracking and Command Station Blossom Point*	0/0	0/265	22/23
Florida			
Marine Corrosion Facility Key West	Tenant		
California			
NRL Monterey Monterey*	Tenant		
Mississippi			
Stennis Space Center Bay St. Louis*	Tenant		
Alabama			
Ex-USS *Shadwell* (LSD-15) Mobile Bay	Tenant		
Decommissioned 457-ft vessel used for fire research			

PROPERTY

Land: 824 acres

Buildings:
RDT&E 3,183,094 ft^2
Administrative 249,121 ft^2
Other 266,749 ft^2

Replacement Costs:
Buildings Plant Replacement Value (PRV)[1] $1,252.0 million
Equipment Costs[2] $523.7 million

[1] Per DON Facilities Asset Data System standard cost factors.
[2] NRL Accountable Property Acquisition Costs
*See maps in the General Information section (page 131).

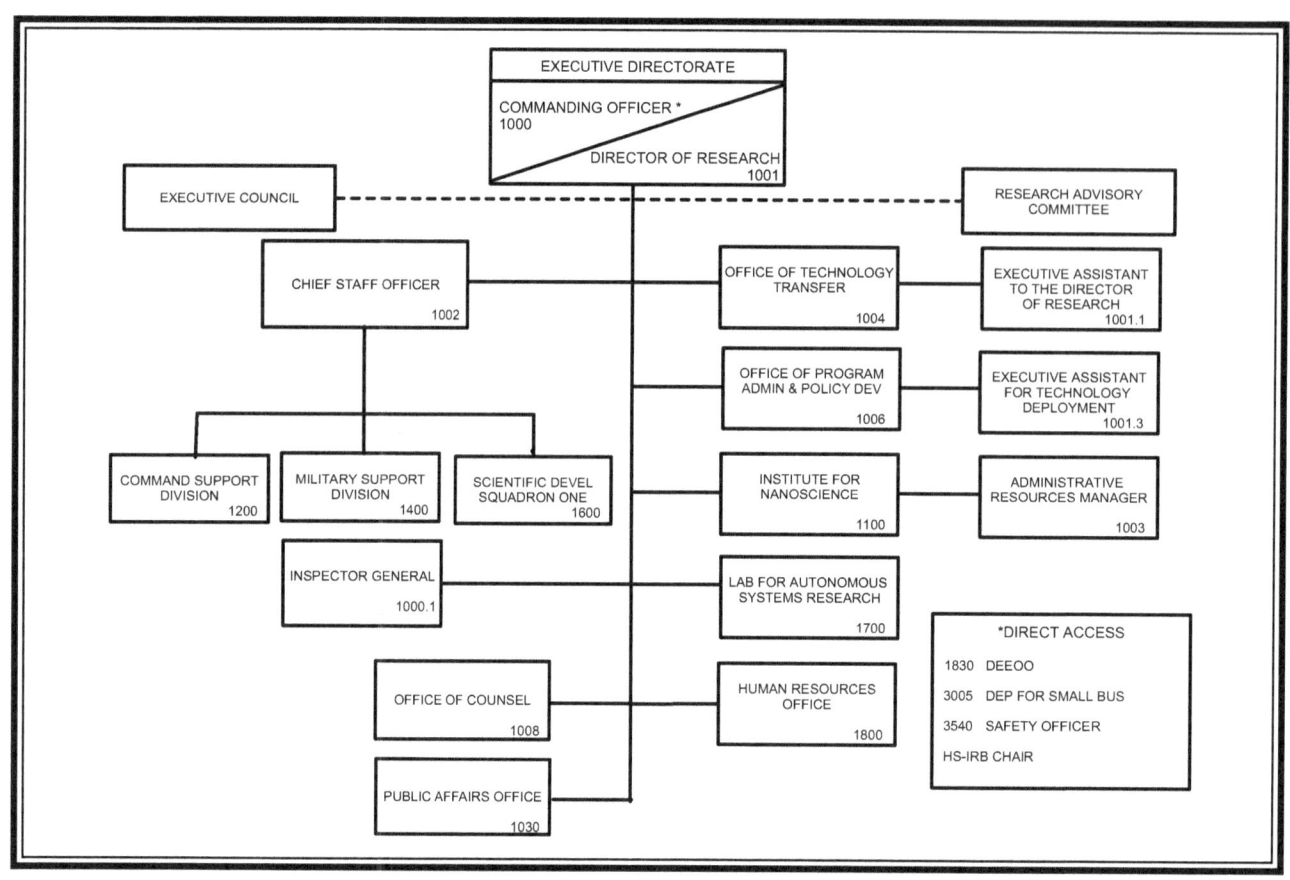

Key Personnel

Title	Code
Commanding Officer	1000
Director of Research	1001
Executive Assistant to the Director of Research	1001.1
Head, Strategic Workforce Planning	1001.2
Executive Assistant for Technology Deployment	1001.3
NRL Historian	1001.15
Chief Staff Officer/Inspector General	1002/1000.1
Command Management Review	1000.12
Head, Office of Technology Transfer	1004
Head, Office of Program Administration and Policy Development	1006
Head, Office of Counsel	1008
Head, Public Affairs Office	1030
Director, Institute for Nanoscience	1100
Head, Command Support Division	1200
Head, Military Support Division	1400
Commanding Officer, Scientific Development Squadron One (VXS-1)	1600
Director, Laboratory for Autonomous Systems Research	1700
Director, Human Resources Office	1800
Deputy Equal Employment Opportunity Officer	1830
Deputy for Small Business	3005
Head, Safety Branch	3540

EXECUTIVE DIRECTORATE

Code 1000 and Code 1001

The Commanding Officer (Code 1000) and the Director of Research (Code 1001) share executive responsibility for the management of the Naval Research Laboratory. In accordance with Navy requirements, the Commanding Officer is responsible for the overall management of the Laboratory and exercises the usual functions of command including compliance with legal and regulatory requirements, liaison with other military activities, and the general supervision of the quality, timeliness, and effectiveness of the technical work and of the support services.

The Commanding Officer delegates line authority and assigns responsibility to the Director of Research for the Laboratory's technical program, its planning, conduct, and staffing; evaluation of the technical competence of personnel; liaison with the scientific community; selection of subordinate technical personnel; exchange of technical information; and the effective execution of the NRL mission.

Within the limits of Navy regulations, the Commanding Officer and the Director of Research share authority and responsibility for the internal management of the Laboratory. The Commanding Officer retains all authority and responsibility specifically assigned to him by higher authority.

The mission of the Laboratory is carried out by three science and technology directorates and the Naval Center for Space Technology, supported by the Business Operations Directorate and the Executive Directorate. In addition, the Laboratory's operating staffs provide assistance in their special fields to the Commanding Officer and to the Director of Research. The operating staffs are listed on the following pages of this publication.

Captain **Anthony J. Ferrari** is a native of Queens, NewYork, and was raised in the New York/New Jersey area. Upon graduation from Delran High School in 1982, he joined the Navy and attended the Naval Academy Preparatory School in Newport, Rhode Island. In 1983, he received an appointment to the United States Naval Academy and graduated in 1987 with a B.S. degree in oceanography and physics. Upon commissioning, he attended undergraduate flight training and was winged as a Naval Flight Officer in 1988. His next set of orders sent him to Whidbey Island, Washington, and Fleet Replacement Squadron 128 (VA-128), where he completed Bombardier/Navigator training in 1990 and joined the "Milestones" of VA-196. During his tour with VA-196, he accumulated over 1,000 hours in the A-6 Intruder and flew missions in support of Operation Desert Shield.

In 1993, he was selected for U.S. Naval Test Pilot School and graduated in the summer of 1994 with class 105. As a Flight Test Officer, he was assigned to VX-23 in Patuxent River, Maryland, and worked on various test projects supporting Carrier Aviation and Weapons testing. When the A-6 Intruder was faithfully retired, he transitioned to the F-14 community and served on the staff of CVW-17 as the Air Wing Strike Operations Officer, completing two Mediterranean deployments from 1997 to 1999. Following a brief training syllabus at VF-101, he reported to the "World Famous Pukin' Dogs" (VF-143) and served as the Safety and Operations Officer.

Upon completion of his department head tour, he was then assigned as the Officer in Charge and Chief Operational Test Director of the VX-9 detachment, Point Mugu, California. This tour was followed by a second tour in Patuxent River, joining NAVAIR as the PMA-241 class desk officer, and principal deputy Program Manager. During this tour, he transitioned to the Aviation Engineering Duty Officer (AEDO) community, was selected as an Acquisition Professional (AP), and received an M.S. degree in systems engineering at Johns Hopkins University.

After leaving NAVAIR, he was assigned as the Naval Aviation Depot Requirements Officer, Fleet Readiness Division (OPNAV N43) in the Office of the Chief of Naval Operations at Washington, DC. This was followed by a tour with the Naval Personnel Command as the Head Detailer for the Aerospace Engineering and Maintenance Communities.

Selected for Major Command in 2008, he proudly served as the Deputy Director and Director of PMR-51, the Navy's Low Observable/Counter Low Observable Technology, Policy and Advanced Project office from December 2008 through August 2012.

Captain Ferrari has been awarded the Legion of Merit, Meritorious Service Medal (four awards), Navy and Marine Corps Commendation Medal (four awards) and the Navy and Marine Corps Achievement Medal (three awards), in addition to numerous campaign and unit awards.

Dr. John A. Montgomery joined the Naval Research Laboratory in 1968 as a research physicist in the Advanced Techniques Branch of the Electronic Warfare Division, where he conducted research on a wide range of Electronic Warfare (EW) topics. In 1980, he was selected to head the Off-Board Countermeasures Branch. In May 1985, he was appointed to the Senior Executive Service and was selected as Superintendent of the Tactical Electronic Warfare Division. He has been responsible for numerous systems that have been developed/approved for operational use by the Navy and other services. He has had great impact through the application of advanced technologies to solve unusual or severe operational deficiencies noted during world crises, most recently in Afghanistan, Iraq, and for Homeland Defense and in the Pacific theater. Dr. Montgomery has accumulated 43 years of civilian service to-date at the Naval Research Laboratory.

Dr. Montgomery received the Department of Defense Distinguished Civilian Service Award in 2001. He was recognized by the Department of the Navy Distinguished Civilian Service Award in 1999 and by the Department of the Navy Meritorious Civilian Service Award in 1986. As a member of the Senior Executive Service, he received the Presidential Rank Award of Distinguished Executive in 1991 and again in 2002, and the Presidential Rank Award of Meritorious Executive in 1988, 1999 and again in 2007. He also received the 1997 Dr. Arthur E. Bisson Prize for Naval Technology Achievement, awarded by the Chief of Naval Research in 1998. Further, he has received the Association of Old Crows (Electronic Defense Association) Joint Services Award in 1993. He was an NRL Edison Scholar, and is a member of Sigma Xi. He served as the U.S. National Leader of The Technical Cooperation Program's multinational Group on Electronic Warfare from 1987 to 2002, and served as its Executive Chairman. In 2006, Dr. Montgomery received the Laboratory Director of the Year award from the Federal Laboratory Consortium for Technology Transfer, and in 2011, he received the Roger W. Jones Award for Executive Leadership from American University's School of Public Affairs.

Dr. Montgomery received his bachelor's of science degree in physics from North Texas State University in 1967 and his master's degree, also in physics, in 1969. He received his PhD in physics from the Catholic University of America in 1982. As Director of Research at the Naval Research Laboratory, Dr. Montgomery oversees research and development programs with expenditures of approximately $1.2 billion per year.

The Executive Council consists of executive, management, and administrative personnel. Executive Council members include the following:

Commanding Officer, Chairperson
Director of Research
Executive Assistant to the Director of Research
Associate Directors of Research
Chief Staff Officer
Director, Naval Center for Space Technology
Associate Director, Naval Center for Space Technology
Heads of Divisions
Director, Laboratories for Computational Physics and Fluid Dynamics
Director, Center for Bio/Molecular Science and Engineering
Director, Human Resources Office
Public Affairs Officer
Deputy Equal Employment Opportunity Officer
Administrative Resources Manager
Head, Office of Program Administration and Policy Development
Safety Officer
Head, Office of Counsel
Head, Office of Technology Transfer
Head, Management Information Systems Staff
Head, Office of Research Support Services
Representative, Administrative Advisory Council
Director, Institute for Nanoscience
Director, Laboratory for Autonomous Systems Research

The Research Advisory Committee advises the Commanding Officer and the Director of Research on scientific programs and the administration of the Laboratory. The committee assists in planning the long-range scientific program, coordinating the scientific work, reviewing the budget, accepting or modifying problems, considering personnel actions, and initiating such studies as may be necessary or desirable. The membership consists of the following:

    Director of Research, Chairperson
    Commanding Officer
    Associate Directors of Research
    Chief Staff Officer (Observer)

Chief Staff Officer/Inspector General
Code 1002/1000.1

CAPT K. SZCZUBLEWSKI, USN

The Chief Staff Officer serves as the Deputy to the Commanding Officer and acts for the Commanding Officer in his absence. The Command Support Division (Code 1200), the Military Support Division (Code 1400), and the Scientific Development Squadron One (VSX-1) (NAS Patuxent River, MD, Code 1600) report directly to the Chief Staff Officer. When directed, the Laboratory's Inspector General investigates, inspects, and/or inquires into matters that affect the operation and efficiency of NRL. These matters include but are not limited to: effectiveness, efficiency, and economy; management practices; and fraud, waste, and abuse. He serves as principal advisor to the Commanding Officer on all inspection matters and audits and is the principal point of contact and liaison with all agencies outside NRL.

Public Affairs Officer
Code 1030

MR. R.L. THOMPSON

The Public Affairs Officer (PAO) advises the Commanding Officer and Director of Research on public affairs matters, including external and internal relations and community outreach, and serves as the Commanding Officer's principal assistant in the area of public affairs. To do this, the PAO plans and directs a program of public information dissemination on official NRL activities. The PAO coordinates responses to requests from the news media and the public for unclassified information or materials dealing with the Laboratory, coordinates participation in community relations activities, and directs the internal information programs. The PAO is also responsible for coordinating all actions within the Laboratory that respond to requirements of the Freedom of Information Act (FOIA).

Deputy Equal Employment Opportunity Officer
Code 1830

MS. L.L. HILL

The Deputy Equal Employment Opportunity Officer (DEEOO) is the EEO program manager and the advisor to the Commanding Officer on all EEO matters. The DEEOO manages the discrimination complaint process and directs the Laboratory's affirmative action plans and special emphasis programs (Federal Women's, Hispanic Employment, African American Employment, Asian-Pacific Islanders, American Indian Employment, Individuals with Disabilities, including Disabled Veterans). The DEEOO recruits quality candidates for those areas when underrepresentation exists. Duties also include reviewing, coordinating, and monitoring implementation of EEO policies and developing local guidance, directives, and implementation procedures for the EEO programs.

Office of Technology Transfer

Code 1004

Basic Responsibilities

The Technology Transfer Office (TTO) is responsible for NRL's implementation of the Federal Technology Transfer Act of 1986 (Public Law 99-502). The law requires the transfer of Government innovative technologies to industry for commercialization as products and services for public benefit. TTO negotiates Cooperative Research and Development Agreements (CRADAs) under which NRL investigators collaborate with investigators from industry, academia, state or local governments, or other Federal agencies to develop NRL technologies for government and/or commercial use. It markets NRL's patented inventions, negotiates patent license agreements under which the Navy grants a licensee the right to make, use, and sell NRL inventions (in exchange for receiving licensing fees and a percentage of sales), and enforces licenses to assure diligence in commercialization efforts.

Personnel: 6 full-time civilian; 1 SCEP student, 1 STEP student

Key Personnel

Title	Code
Head, Technology Transfer	1004
Sr. Licensing Associate	1004
Sr. Licensing Associate	1004
Social Media Marketing Associate	1004
Licensing Associate	1004
Management Analyst	1004
Administrative Assistant (SCEP)	1004
Administrative Assistant (STEP)	1004

Point of contact: Code 1004, (202) 767-7229

Office of Program Administration and Policy Development

Code 1006

Basic Responsibilities

The Office of Program Administration and Policy Development provides managerial, technical, and administrative support to the Director of Research (DOR) in such areas as program and policy development, intra-Navy and inter-Service Science and Technology (S&T) program coordination; liaison with other Navy, DoD, and government activities on matters of mutual concern; and support to the Executive Directorate in planning and directing NRL's S&T (6.1, 6.2) program. Specific functions include: monitoring and providing background information on technical and policy matters that come under the purview of the DOR; representing NRL, ONR, and/or the Navy on tri-Service or DoD-wide coordination matters; performing special studies or chairing ad hoc study groups regarding program decisions or policy positions; performing special studies involving major NRL programs and resource issues; providing administrative support in the areas of personnel, budget, facilities, equipment, and security; providing executive management information and analyses for various aspects of the S&T program effort; coordinating VIP visits to NRL; managing the NRL directives system; administering the NRL response to Congressional requests; maintaining the NRL R&D achievements file; developing the S&T guidance for monitoring and reporting the NRL S&T program; administering NRL's various postdoctoral fellowship programs; and managing the Facility Modernization Program.

Personnel: 14 full-time civilian

Key Personnel

Title	Code
Head, Office of Program Administration and Policy Development	1006
Head, Program Administration Staff	1006.1
VIP Coordinator/Protocol Officer	1006.2
Head, Executive Management & Policy Development Staff	1006.3
Directives	1006.31
Head, NRL Facilities Staff	1006.4
Special Assistant	1006.6

Point of contact: Code 1006.2, (202) 767-3370

Office of Counsel

Code 1008

Basic Responsibilities

The Office of Counsel is responsible for providing legal services to NRL's management in all areas of general, administrative, intellectual property, and technology transfer law. The Office reviews all procurement-related actions; reviews NRL scientific papers prior to publication; prepares patent applications and prosecutes the applications through the Patent and Trademark Office; defends against contract protests, other contract litigation, and personnel cases; and advises on other legal matters relating to technology transfer, personnel, fiscal, and environmental law.

NRL Counsel also serves as legal advisor to the Commanding Officer and Director of Research.

Personnel: 30 full-time civilian

Key Personnel

Title	Code
Head, Office of Counsel	1008
Associate Counsel/General Law	1008.1
Associate Counsel/Intellectual Property	1008.2
Associate Counsel/SSC Legal Matters	1008.3

Point of contact: Code 1008.1, (202) 767-7606

Code 1100
Staff Activity Areas

• Interdisciplinary nanoscience that enables:
  Low-power, high-speed electronics
  Lightweight, high-strength materials
  Highly sensitive molecular sensors
  Efficient energy generation and storage

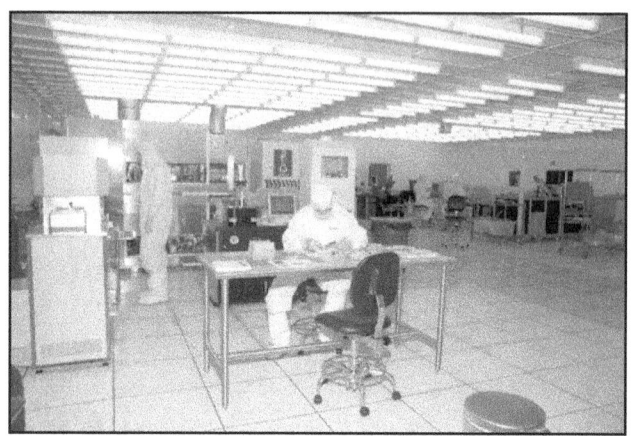

NRL researchers working in the Class 100 clean room in the Institute for Nanoscience.

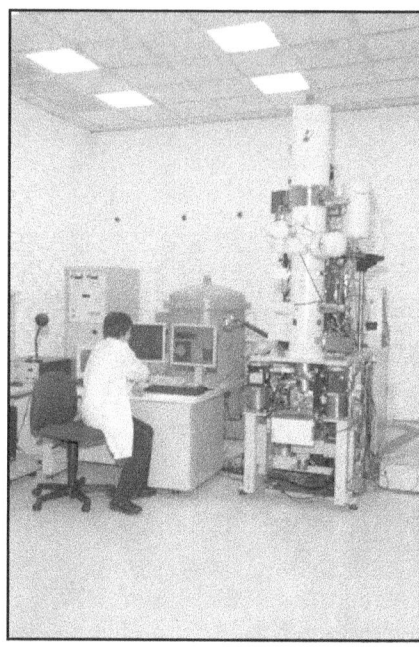

Transmission electron microscope located in one of the Institute for Nanoscience's environmentally controlled laboratories.

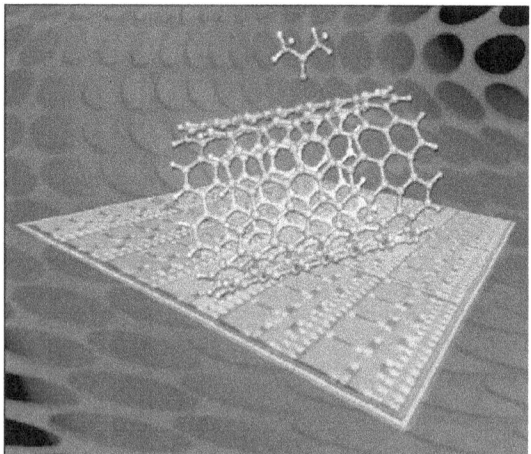

Wafer of carbon nanotube chemical sensors fabricated in the Institute for Nanoscience clean room.

Code 1100

Basic Responsibilities

The Institute for Nanoscience has two primary responsibilities: to administer an interdisciplinary research program in nanoscience and to provide NRL scientists with high-quality laboratory space and state-of-the-art nanofabrication facilities.

The mission of the research program is to conduct highly innovative, interdisciplinary research at the intersections of the fields of materials, electronics, and biology in the nanometer size domain. The Institute exploits the broad multidisciplinary character of NRL to bring together scientists and engineers with disparate training and backgrounds to attack common goals at the intersection of their respective fields at this length scale. The Institute's S&T programs provide the Navy and DoD with scientific leadership in this complex, emerging area and help to identify opportunities for advances in future defense technology.

The Institute also operates a nanoscience research building containing nanofabrication facilities and environmentally controlled measurement laboratories. The central core of the building, a 5000 sq ft Class 100 clean room, has been outfitted with the newest tools to permit nanofabrication, measurement, and testing of devices. In addition to the clean room facility, the building also contains 5000 square feet of controlled-environment laboratory space, which is available to NRL researchers whose experiments are sufficiently demanding to require this space. There are 12 of these laboratories within the building. They provide shielding from electromagnetic interference, and very low floor vibration and acoustic levels. Eight of the laboratories control the temperature to within $\pm 0.5\,°C$ and four to within $\pm 0.1\,°C$.

Personnel: 3.5 full-time civilian

Key Personnel

Title	Code
Director, Institute for Nanoscience	1100
Position Assistant	1100
Facilities Manager	1100
Facilities Manager	1100

Point of Contact: Code 1100, (202) 767-1804

Code 1200
Staff Activity Areas

• Security

Incoming visitor reception area

Security monitoring

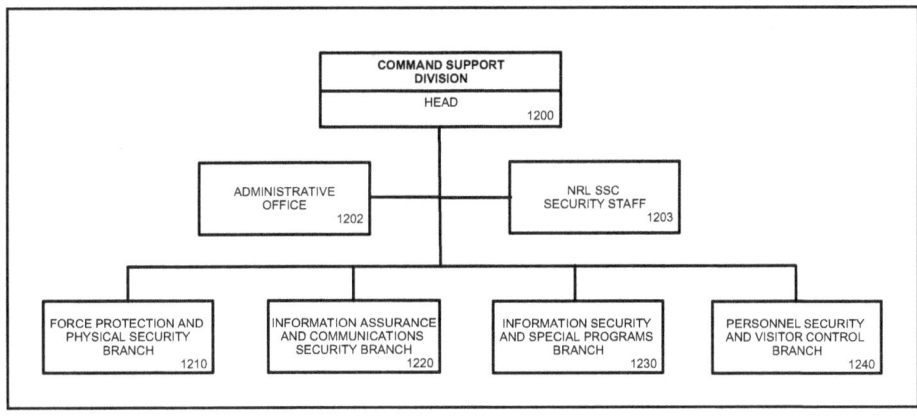

Basic Responsibilities

The Command Support Division is responsible for NRL security policy, management, and enforcement. The Division Head is the NRL Security Manager. The primary areas of security are: information assurance, information security, personnel security, industrial security, classification management, public release, foreign disclosure, physical security, force protection, antiterrorism, operations security, special security programs, and communications security. Provides security education across all security disciplines. Conducts local inspections for compliance with current internal and external policies. Provides advice and guidance to senior NRL management concerning the security posture of the Command. Provides administrative budget support to the Military Support Division (Code 1400) and Scientific Development Squadron One (VXS-1, Code 1600).

Personnel: 50 full-time civilian

Key Personnel

Title	Code
Head, Command Support Division	1200
Administrative Officer	1202
Head, Stennis Space Center Security Staff	1203
Head, Force Protection and Physical Security Branch	1210
Head, Information Assurance and Communications Security Branch	1220
Head, Information Security and Special Programs Branch	1230
Head, Personnel Security and Visitor Control Branch	1240

Point of contact: Code 1202, (202) 767-6987

Code 1400
Staff Activity Areas

- Operations
- Administrative Operations

P-3 airborne research platform

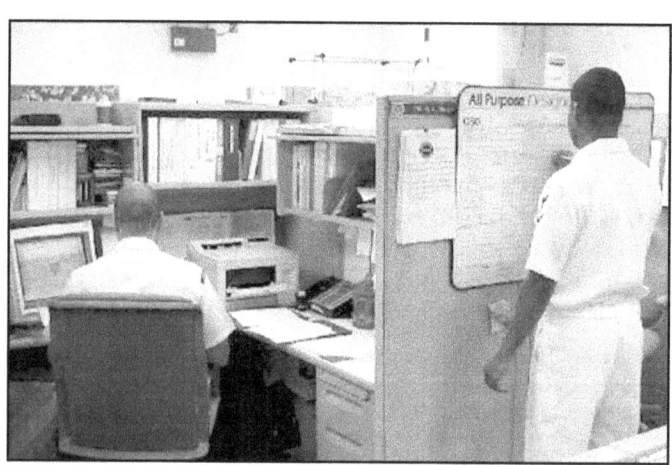

Administration

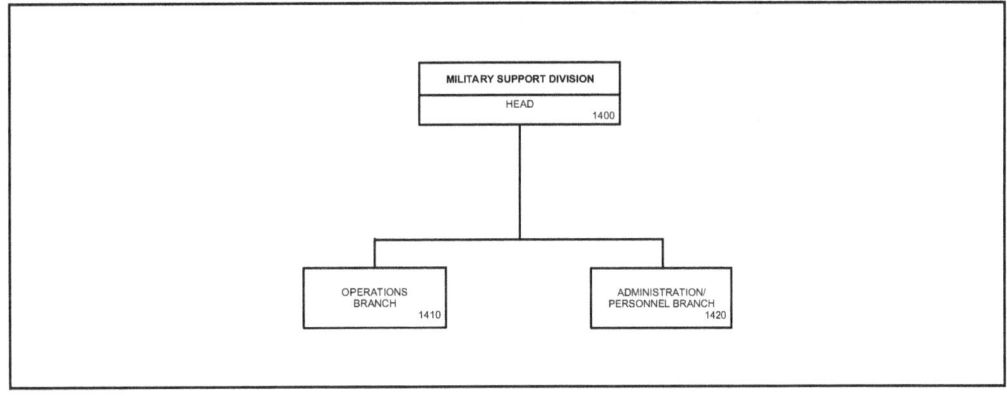

Basic Responsibilities

The Military Support Division provides military operational and administrative services to NRL.

The Operations Branch assists NRL research directorates in planning and executing project flight missions, develops deployment schedules and military operational and training objectives, and coordinates the Research Reserve Program within NRL.

The Military Administration Branch is responsible for the coordination and efficient functioning of all military administrative operations for NRL (including site detachments). These duties specifically include: personnel actions, maintenance of personnel records, performance evaluations, awards and training; advising the Chief Staff Officer on manpower matters and organization issues; and preparing and administering the military operational budget.

Personnel: 1 full-time civilian; 7 military

Key Personnel

Title	Code
Head, Military Support Division	1400
Project Officer	1410
Project Officer	1410
Project Officer	1410
Administrative Officer	1420

Point of contact: Code 1420, (202) 767-7632

Code 1600
Staff Activity Areas

- Operations
- Administrative Operations
- Aircraft Maintenance
- Safety/NATOPS

VXS-1 maintains two RC-12 aircraft dedicated to airborne research. They are smaller, more cost-efficient alternatives to the P-3 Orion. Each aircraft is outfitted with a research electrical load center and has a roll-on roll-off capability which enables it to be equipped with project stations. The RC-12s can support a broad spectrum of project configurations.

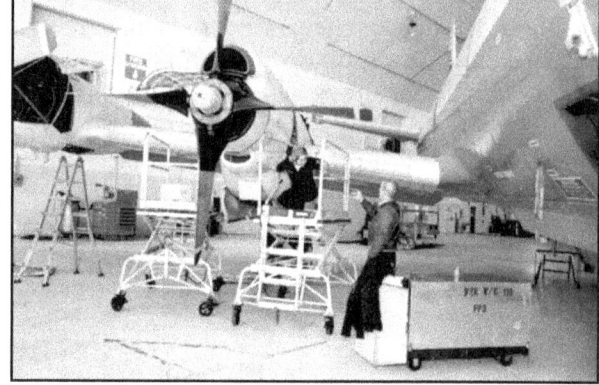

Aircraft maintenance

P-3 airborne research platform

Scientific Development Squadron One hangar

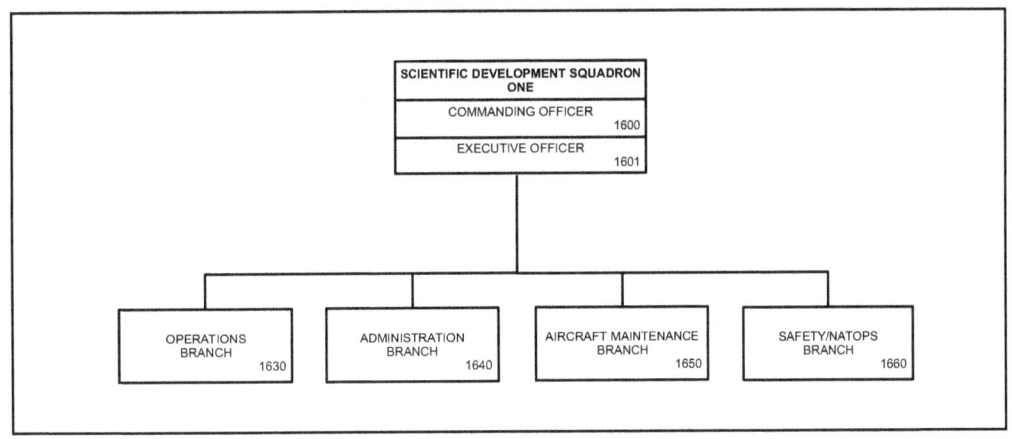

Basic Responsibilities

The Scientific Development Squadron ONE (VXS-1) located at NAS Patuxent River, Maryland, operates and maintains five uniquely configured P-3 Orion aircraft and two C-12 aircraft. The men and women of the squadron provide the Naval Research Laboratory with airborne research platforms, conducting flights world-wide in support of a broad spectrum of projects and experiments. These include magnetic variation mapping, hydroacoustic research, bathymetry, electronic countermeasures, gravity mapping, and radar research. The squadron annually logs approximately 1000 flight hours, and in its 47 years, Scientific Development Squadron ONE (VXS-1) has amassed 69,000 hours of mishap-free flying.

Personnel: 3 full-time civilian; 70 military; 9 full-time contractors

Key Personnel

Title	Code
Commanding Officer, VXS-1	1600
Executive Officer	1601
Senior Enlisted Advisor	1600.2
Executive Secretary	1600.4
Operations Officer	1630
Administrative Officer/Public Affairs Officer	1640
Maintenance Officer	1650
Assistant Maintenance Officer	1650.1
Maintenance/Material Control Officer	1650.2
Safety/NATOPS Officer	1660

Point of contact: Code 1600.4, (301) 342-3526; DSN 342-3526

Laboratory for Autonomous Systems Research

Code 1700
Staff Activity Areas

Multidisciplinary research, development, and integration in autonomous systems, including:
- Software for intelligent autonomy
- Novel human-systems interaction technology
- Mobility and platforms
- Sensor systems
- Power and energy systems
- Networking and communications
- Trust and assurance

The Laboratory for Autonomous Systems Research integrates S&T components into research prototype systems.

Because autonomous systems are not just vehicles, the building contains a number of human-system interaction labs to develop automated decision support tools and address critical communications and network issues.

The Reconfigurable High Bay allows operation of small air vehicles as well as ground vehicles.

Code 1700

Basic Responsibilities

The Laboratory for Autonomous Systems Research provides specialized facilities to support highly innovative, interdisciplinary research in autonomous systems, including software for intelligent autonomy, sensor systems, power and energy systems, human-systems interaction, networking and communications, and platforms and mobility. The Laboratory capitalizes on the broad multidisciplinary character of NRL, bringing together scientists and engineers with disparate training and backgrounds to advance the state of the art in autonomous systems at the intersection of their respective fields. The Laboratory provides unique facilities and simulated environments (littoral, desert, tropical) and instrumented reconfigurable high bay spaces to support integration of science and technology components into research prototype systems. The objective of the laboratory is to enable Naval and DoD scientific leadership in this complex, emerging area and to identify opportunities for advances in future defense technology.

The facility includes a Reconfigurable Prototyping High Bay that allows real-time, accurate tracking of many entities (vehicles and humans) for experimental ground truth. Small UAVs and ground vehicles can simultaneously operate within the large high bay, which is viewable from four adjacent Human-System Interaction labs. The Tropical High Bay emulates a rainforest with appropriate terrain and plants, and includes flowing water features. An outdoor Highland Forest provides an additional forest environment, and also includes interesting water and terrain features. The Desert High Bay provides a simulated desert environment featuring as sand pit, natural rock walls, and appropriate lighting and wind. The Littoral High Bay provides a simulated coastal environment featuring sediment tanks, large pool with a sloping floor, and small flow tanks. In addition to the environmental high bays, the facility also has a Power and Energy Laboratory, a Sensor Laboratory, and a mechanical and electrical shop.

The facility is open to use by all NRL scientists contributing to the science and technology of autonomous systems and will host many NRL scientists as needed.

Personnel: 1.5 full-time civilian

Key Personnel

Title	Code
Director, Laboratory for Autonomous Systems Research	1700
Facilities Manager	1700

Point of contact: Code 1700, (202) 767-2684

Human Resources Office

Code 1800
Staff Activity Areas

- Personnel Operations (Staffing and Classification)
- Employee Relations
- Employee Development
- Equal Employment Opportunity and Manpower
- Compensation, Reports, and Demonstration Project
- Information Technology and Reports

Personnel Operations Branch

Diversity and Employee Recognition Branch

Employee Relations Branch

Employee Development and Management Branch

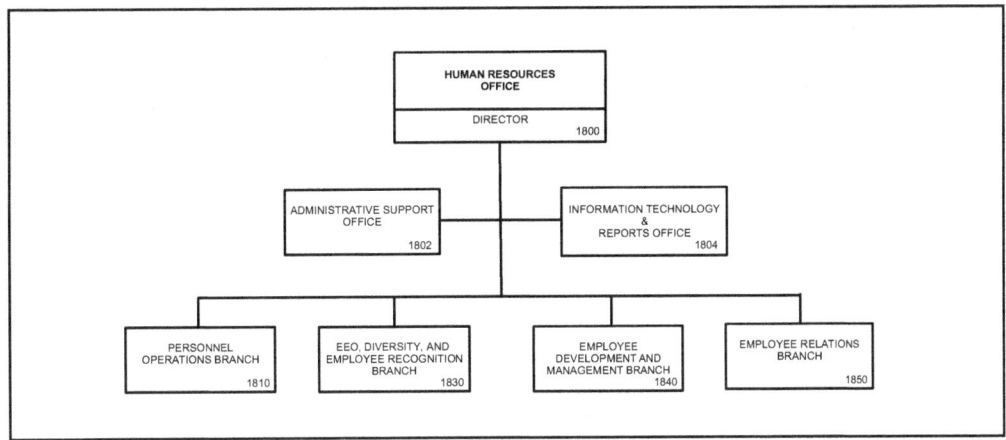

Basic Responsibilities

The Human Resources Office (HRO) provides civilian personnel, manpower, and Equal Employment Opportunity (EEO) services to the Naval Research Laboratory. The Human Resources Program provides the full range of operating civilian personnel management in the staffing and placement, position classification, employee relations, labor relations, employee development, EEO functional areas, manpower management, and morale, welfare, and recreation programs.

The HRO at NRL's main site in Washington, DC, services approximately 2,500 employees and provides a centralized capability to perform managerial, service, and advisory functions in support of field office operations. These include issuing policy and procedural directives; developing, designing, and maintaining automated systems; and monitoring and evaluating product effectiveness to develop and maintain efficient, cost-effective, service-oriented methods.

Personnel: 30 full-time civilian

Key Personnel

Title	Code
Director, Human Resources Office	1800
Administrative Officer	1802
Head, Information Technology and Reports Office	1804
Head, Personnel Operations Branch	1810
Head, EEO, Diversity, and Employee Recognition Branch	1830
Head, Employee Development and Management Branch	1840
Head, Employee Relations Branch	1850

Point of contact: Code 1802, (202) 404-2797

Ruth H. Hooker Research Library

Code 5596

Basic Responsibilities

NRL's Ruth H. Hooker Research Library supports NRL and ONR scientists in conducting their research by making a comprehensive collection of the most relevant scholarly information available and useable; by providing direct reference and research support; by capturing and organizing the NRL research portfolio; and by creating, customizing, and deploying a state-of-the-art digital library. Traditional library resources include extensive technical report, book, and journal collections dating back to the 1800s housed within a centrally located research facility that is staffed by subject specialists and information professionals. The collections include 44,000 books; 80,000 digital books; 80,000 bound historical journal volumes; more than 3,500 current journal subscriptions; and approximately 2 million technical reports in paper, microfiche, or digital format (classified and unclassified). Research Library staff members provide advanced information consulting; literature searches against all major online databases including classified databases; circulation of materials from the collection including classified literature up to the Secret level; and retrieval of articles, reports, proceedings, or documents through our interlibrary loan and document delivery network. The digital library provides desktop access to thousands of journals, books, proceedings, reports, databases, and reference sources.

Personnel: 21 full-time civilian

Key Personnel

Title	Code
Chief Librarian	5596
Head, Research Reports and Bibliography	5596.3
Library IT Director	5596.2

Point of contact: Code 5596, (202) 767-2357

Business Operations Directorate

BUSINESS OPERATIONS DIRECTORATE

Code 3000

The Business Operations Directorate provides executive management, policy development, and program administration for business programs needed to support the activities of the scientific directorates. This support is in the areas of financial management, supply management, technical information services, contracting, research and development services, and management information systems support.

Mr. D.K. Therning was born in Modesto, California. He graduated from Washington State University with a bachelor's degree in finance in 1983 and earned a master's degree in business administration from George Mason University in 1993. Mr. Therning has accumulated extensive experience in the financial business management of research, development, test, and evaluation (RDT&E) activities within the Department of the Navy (DON) beginning at the Naval Weapons Center, China Lake, California, where he served as a budget analyst in the Public Works Department and then in the Weapons Department. In 1984, he became the Financial Management Advisor to the Ordnance Systems Department. In 1985, under the auspices of the Naval Scientist Training and Exchange Program, he was selected for a one-year assignment in the Office of the Director of Naval Laboratories (DNL), Washington, DC. He remained on the DNL staff as a budget analyst until 1987, when he was appointed Budget Officer of the DNL's seven Navy Industrial Fund R&D laboratories.

As the DON reorganized the R&D laboratories and T&E activities, Mr. Therning oversaw the financial reorganization of the DNL labs with other activities into the Naval warfare centers. Upon the disestablishment of DNL, Mr. Therning remained in the Space and Naval Warfare Systems Command as the Director of the Defense Business Operations Fund (DBOF) Resources Management Division, with collateral duty as the Financial Manager of the Naval Command, Control, and Ocean Surveillance Center (NCCOSC). During this time, he managed the conversion of nine appropriated fund engineering activities to DBOF and the financial consolidation of these activities with NCCOSC.

In 1995, Mr. Therning served as Head of the Revolving Funds Branch of the Office of the Assistant Secretary of the Navy (Financial Management and Controller), where he was responsible for the budget formulation and execution processes of all DON DBOF activities, which includes the RDT&E activities, shipyards, aviation depots, ordnance centers, and supply centers.

Mr. Therning was appointed Head, Financial Management Division/Comptroller of NRL in July 1996. In October 1996, in addition to leading the Financial Management Division, he assumed responsibilities for the Management Information Systems office. In January 1999, as an additional duty to his role as Comptroller, Mr. Therning was appointed to the newly established position of Deputy Associate Director of Research for Business Operations to assist in the management and administration of the Business Operations Directorate.

Mr. Therning was Acting Associate Director of Research for Business Operations from April 1999 until March 2000, when he was appointed the Associate Director of Research for Business Operations.

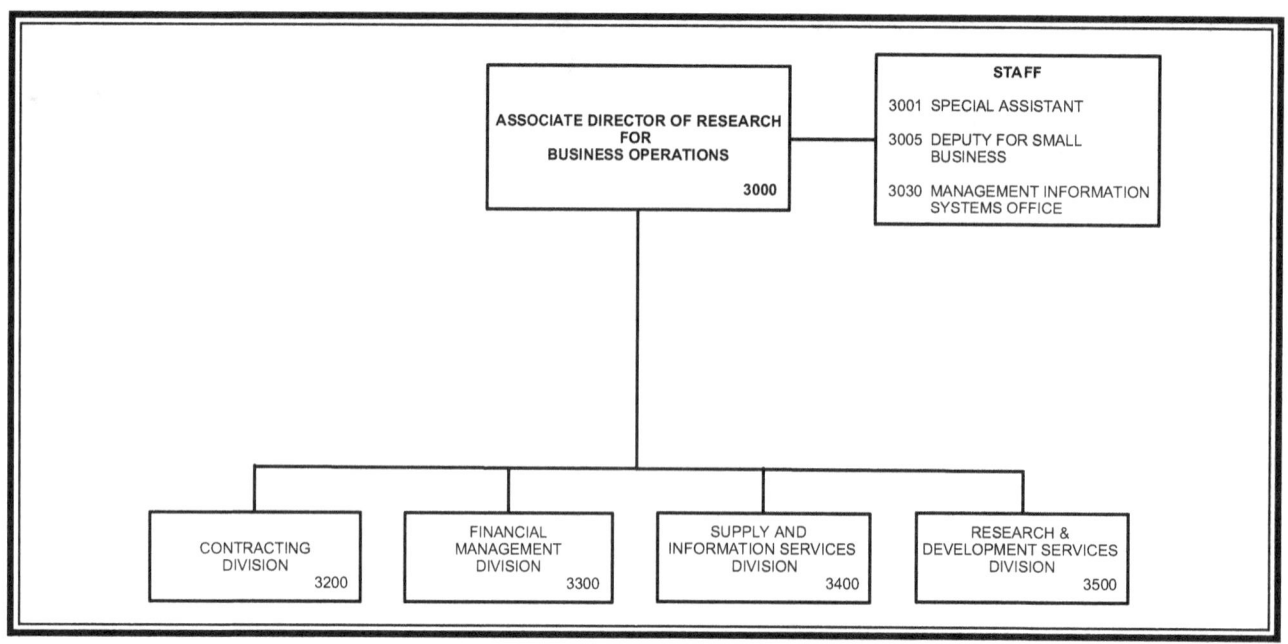

Key Personnel

Title	Code
Associate Director of Research for Business Operations	3000
Special Assistant	3001
Deputy for Small Business	3005
Head, Management Information Systems Office	3030
Head, Contracting Division	3200
Head, Financial Management Division	3300
Head, Supply and Information Services Division	3400
Director, Research and Development Services Division	3500

Point of contact: Code 3000A, (202) 404-7461

Contracting Division

Code 3200
Staff Activity Areas

- Advance Acquisition Planning
- Acquisition Strategies
- Acquisition Training
- Contract Negotiations
- Contractual Execution
- Contract Administration
- Acquisition Policy Interpretation and Implementation

Customers are greeted at the receptionist station.

Contracting personnel attend training session.

Procurement Technician reviews contract file.

Specialist and Division Head discuss small business programs.

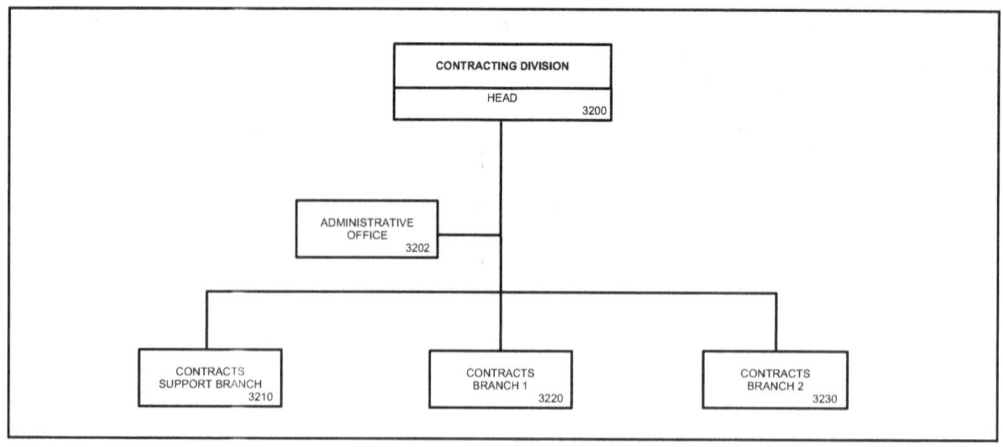

Basic Responsibilities

The Contracting Division is responsible for the acquisition of major research and development materials, services, and facilities where the value is in excess of $100,000. It also maintains liaison with the ONR Procurement Directorate on procurement matters involving NRL. Specific functions include: providing consultant and advisory services to NRL division personnel on acquisition strategy, contractual adequacy of specifications, and potential sources; reviewing procurement requests for accuracy and completeness; initiating and processing solicitations for procurement; awarding contracts; performing contract administration and post-award monitoring of contract terms and conditions, delivery, contract changes, patents, etc., and taking corrective actions as required; providing acquisition-related training to division personnel; and interpreting and implementing acquisition-related Federal, Department of Defense, and Navy regulations.

Personnel: 30 full-time civilian

Key Personnel

Title	Code
Head, Contracting Division	3200
Administrative Officer	3202
Contracts Support Branch	3210
Head, Contracts Branch 1	3220
Head, Contracts Branch 2	3230
Head, Contracts Section, SSC	3235

Point of contact: Code 3202, (202) 767-3749

Financial Management Division

Code 3300
Staff Activity Areas

- Budget
- Reports and Statistics
- Accounting
- Travel Services
- Payroll Liaison

The Budget Branch prepares various financial analyses, reports, and studies in response to external data calls and/or management requests.

The Financial Systems, Reports, and Accounting Branch ensures that NRL's financial system satisfies user requirements and is in compliance with applicable rules and regulations, maintains official accounting records, and coordinates efforts with DFAS to complete payment transactions related to NRL business.

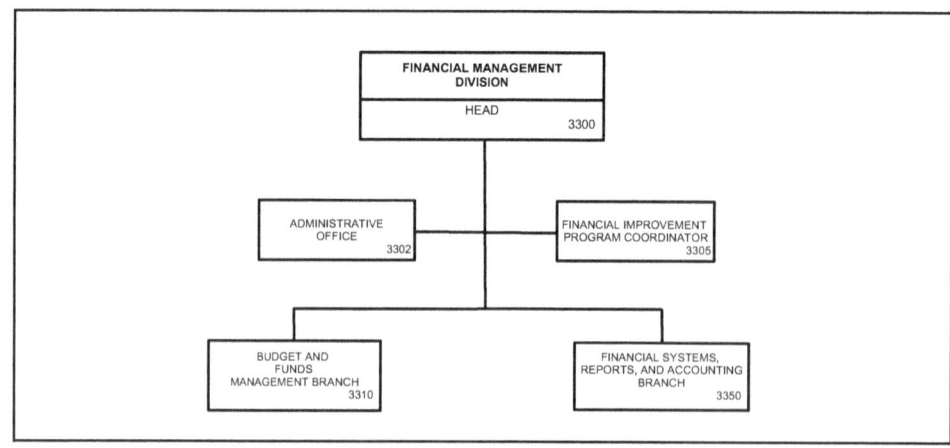

Basic Responsibilities

The Financial Management Division (FMD) develops, coordinates, and maintains an integrated system of financial management that provides the Comptroller, Commanding Officer, Director of Research, and other officials of NRL the information and support needed to fulfill the financial and resource management aspects of their responsibilities. FMD translates the NRL program requirements into the financial plan, formulates the NRL budget, monitors and evaluates performance with the budget plan, and provides recommendations and advice to NRL management for corrective actions or strategic program adjustments. FMD maintains the accounting records of NRL's financial and related resources transactions and prepares reports, financial statements, and other documents in support of NRL management needs and/or to comply with external reporting requirements. FMD provides financial management guidance, policies, advice, and documented procedures to ensure that NRL operates in compliance with Navy and DoD regulations and with economy and efficiency. FMD coordinates efforts with the Defense Finance and Accounting Service (DFAS) to complete payment transactions related to NRL business (e.g., the payment of NRL personnel for payroll and travel expenses and the payment to NRL's contractors and vendors for goods and services purchased by NRL). FMD coordinates Financial Improvement Program efforts to ensure the NRL is ready for an independent financial audit. Additionally, FMD develops, operates, and maintains automated business and management information systems supporting the lab-wide administrative and business processes, including financial management, procurement and contracting, stores and inventory, asset management, human resources, facilities, and security.

Personnel: 68 full-time civilian

Key Personnel

Title	Code
Head, Financial Management Division	3300
Administrative Officer	3302
Financial Improvement and Audit Readiness Coordinator	3305
Head, Budget and Funds Management Branch	3310
Head, Funding Section	3311
Head, Internal Budget Section	3312
Head, Corporate Budget Section	3313
Head, Financial Systems, Reports, and Accounting Branch	3350
Head, Cost Accounting Section	3351
Cost and Analysis Unit	3351.1
Head, Vendor Pay Unit	3351.2
Head, Financial Services Section	3352
Head, Payroll Services Unit	3352.1
Head, Travel Services Unit	3352.2
Head, Accounting Systems and Reports Section	3353
Head, Asset Management and Accounting Section	3354

Point of contact: Code 3302, (202) 767-2950

Code 3400
Staff Activity Areas

- Purchasing
- Technical Information Services
- Customer Support and Program Management
- Material Control
- Administrative Services
- Automated Inventory Management System
- Disposal and Storage

Customers and employee at the Supply store.

Woodworkers prepare boxes for shipping.

Disposal and Storage in Building 49.

Mail clerks sort mail by directorate and file into bins by organizational codes. Mail is bundled and delivered once a day.

The Publications staff discusses design ideas for a new publication.

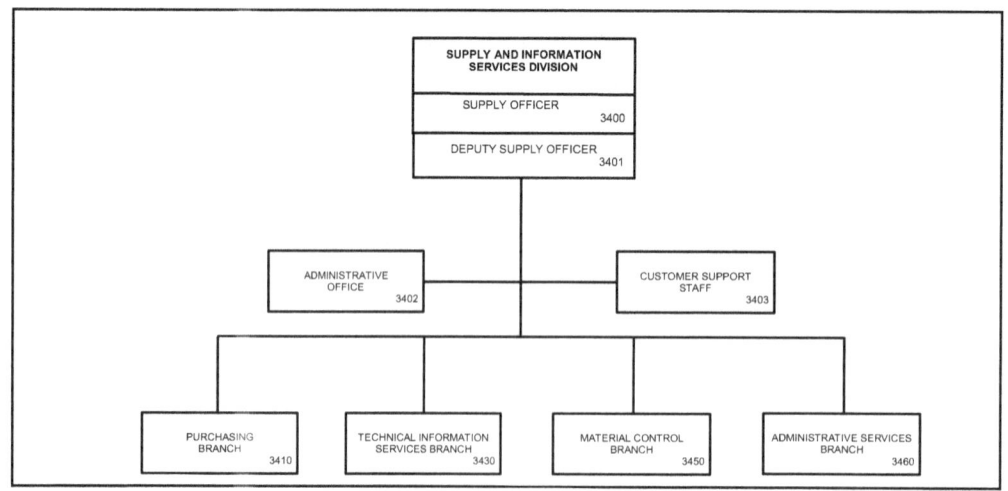

Basic Responsibilities

The Supply and Information Services Division provides the Laboratory and its field activities with contracting, supply management, logistics, administrative, and technical information services. Specific functions include: procuring required equipment, material, and services; receiving, inspecting, storing, and delivering material and equipment; packing, shipping, and traffic management; surveying and disposing of excess and unusable property; operating various supply issue stores and performing stock inventories; providing technical and counseling services for the research directorates in the development of specifications for a complete procurement package; and obtaining and providing guidance in the performance stages of contractual services. Services also include publications, visual information, exhibits, photography, editing, and mailroom services and correspondence management.

Personnel: 102 full-time civilian

Key Personnel

Title	Code
Supply Officer	3400
Deputy Supply Officer	3401
Administrative Officer	3402
Head, Customer Support Staff	3403
Head, Purchasing Branch	3410
Head, Technical Information Services Branch	3430
Head, Material Control Branch	3450
Head, Administrative Services Branch	3460

Point of contact: Code 3402, (202) 404-1701

Code 3500
Staff Activity Areas

- Technical/Support Services
- Production Control
- Shop Services
- Chesapeake Bay Section
- Customer Liaison
- Safety
- Occupational, Safety and Health/Industrial Hygiene
- Explosives Safety
- Health Physics
- Environmental
- Administrative Office
- Telephones
- Facilities Planning and Operations

Safety Office – processing procurement requests for safety equipment

Interstitial hardening furnace

Service Desk – processing service calls

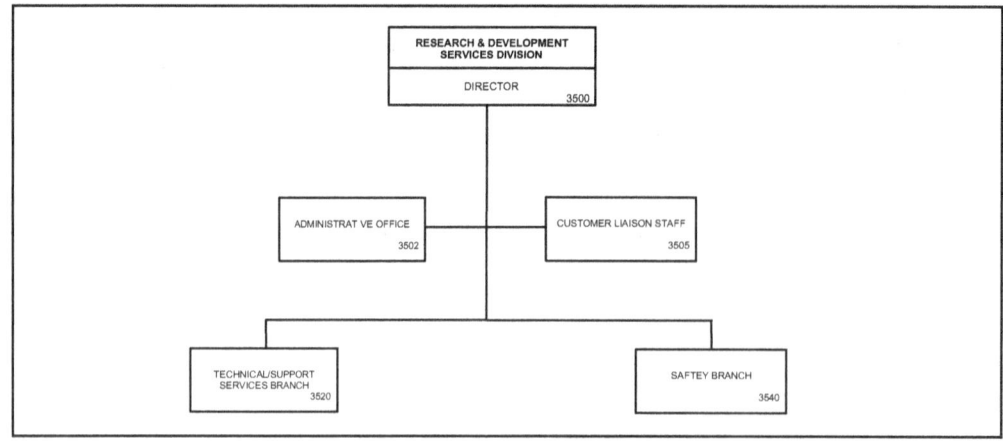

Basic Responsibilities

The Research and Development Services Division is responsible for the physical plant of the Naval Research Laboratory and subordinate field sites. The responsibilities include military construction, engineering, and coordination of construction; facility support services, planning, maintenance/repair/operation of all infrastructure systems; transportation; and occupational safety, health and industrial hygiene, and environmental safety.

The Division provides engineering and technical assistance to research divisions in the installation and operation of critical equipment in support of the research mission.

Personnel: 141 full-time civilian

Key Personnel

Title	Code
Director, Research and Development Services Division	3500
Administrative Officer	3502
Customer Liaison	3505
Head, Technical/Support Services Branch	3520
Head, Engineering Section	3521
Head, Chesapeake Bay Section	3522
Head, Shop Services Section	3523
Head, Production Control Section	3524
Head, Facilities, Planning and Operations Section	3525
Head, Safety Branch	3540
Occupational Safety and Health/Industrial Hygiene Section	3541
Explosives Safety	3542
Health Physics Section	3544
Environmental Section	3546
Environmental Response Unit	3546.1

Point of contact: Code 3502, (202) 404-4312

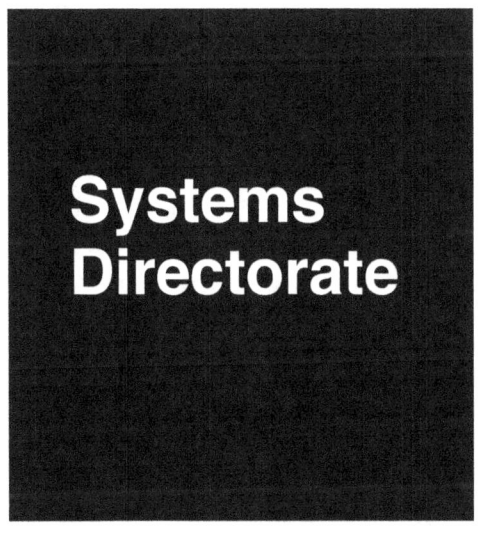

**Systems
Directorate**

SYSTEMS DIRECTORATE

Code 5000

The Systems Directorate applies the tools of basic research, concept exploration, and engineering development to expand operational capabilities and to provide materiel support to Fleet and Marine Corps missions. Emphasis is on technology, devices, systems, and know-how to acquire and move warfighting information and to deny these capabilities to the enemy. Current activities include:

• New and improved radar systems to detect and identify ever smaller targets in the cluttered littoral environment;

• Optical sensors and related materials to extract elusive objects in complex scenes when both processing time and communications bandwidth are limited;

• Unique optics-based sensors for detection of biochemical warfare agents and pollutants, for monitoring structures, and for alternative sensors;

• Advanced electronic support measures techniques for signal detection and identification;

• Electronic warfare systems, techniques, and devices including quick-reaction capabilities;

• Innovative concepts and designs for reduced observables;

• Techniques and devices to disable and/or confuse enemy sensors and information systems;

• Small "intelligent"/autonomous land, sea, or air vehicles to carry sensors, communications relays, or jammers; and

• High performance/high assurance computers with right-the-first-time software and known security characteristics despite commercial off-the-shelf components and connections to public communications media.

Many of these efforts extend from investigations at the frontiers of science to the support of deployed systems in the field, which themselves provide direct feedback and inspiration for applied research and product improvement and/or for quests for new knowledge to expand the available alternatives.

In addition to its wide-ranging multidisciplinary research program, the Directorate provides support to the corporate laboratory in shared resources for high performance computing and networking, technical information collection and distribution, and in coordination of Laboratory-wide efforts in signature technology, counter-signature technology, Theater Missile Defense, and the Naval Science Assistance Program.

Dr. **G.M. Borsuk** is the Associate Director of Research for Systems at the Naval Research Laboratory (NRL) in Washington, DC. In this position he provides executive direction and leadership to four major NRL research divisions that conduct a broad multidisciplinary program of scientific research and advanced technological development in the areas of optics, electromagnetics, information technology, and radar. He is responsible for the conduct and effectiveness of research programs conducted within these divisions and for the overall administration of activities throughout the Systems Directorate. He is also the Focus Area Coordinator for all NRL base programs in electronics science and technology. Prior to this appointment, Dr. Borsuk served for 23 years as the Superintendent of the Electronics Science and Technology Division at NRL where he was responsible for the in-house execution of a multidisciplinary program of basic and applied research in electronic materials and structures, solid state devices, vacuum electronics, and circuits. Dr. Borsuk also serves as the Technical Chair of the DDR&E's Electronic Warfare Technology Task Force (EWTTF). He was the Navy Deputy Program Manager and Technical Director for the now completed DARPA/Tri-Service MIMIC and MAFET Programs. He was the Department of Defense (DoD) technical representative for Electronics to the Wassenaar Arrangement dealing with export control. He has also served as the DoD representative to the President's National Science and Technology Council's Electronic Materials Working Group.

Dr. Borsuk joined the ITT Electro-Physics Laboratory in Columbia, Maryland, as a staff physicist in 1973, where he worked on the application of charge-coupled devices (CCDs) for imaging and signal processing. In 1976 he joined the Westinghouse Advanced Technology Laboratory in Baltimore, Maryland, developing advanced silicon VLSI integrated circuits and performing device physics research. He performed original work in the design and fabrication of CCDs for signal processing and photodetectors for use with acousto-optic signal processors. He headed the Westinghouse VHSIC effort in advanced sub-micron VLSI device technology. Dr. Borsuk was department manager of Solid State Sciences at the Advanced Technology Laboratory when he left Westinghouse in 1983 to join the Naval Research Laboratory as the Superintendent of the Electronics Science and Technology Division.

Dr. Borsuk received a Ph.D. in physics from Georgetown University in Washington, DC, in 1973. He is a Fellow of the IEEE, a member of the American Physical Society, a member of the AVS, and is a member of Sigma Xi. He has 37 technical publications, four patents, and eleven invention disclosures. He is the recipient of four Presidential Rank Senior Executive Awards, the Distinguished, the most recent awarded in 2010. He is also the recipient of the IEEE Frederik Philips Award, the IEEE Harry Diamond Memorial Award, the IEEE Millennium Medal, and an IR-100 Award for his work on high-speed CCDs. Dr. Borsuk also served on the editorial board of the IEEE Proceedings.

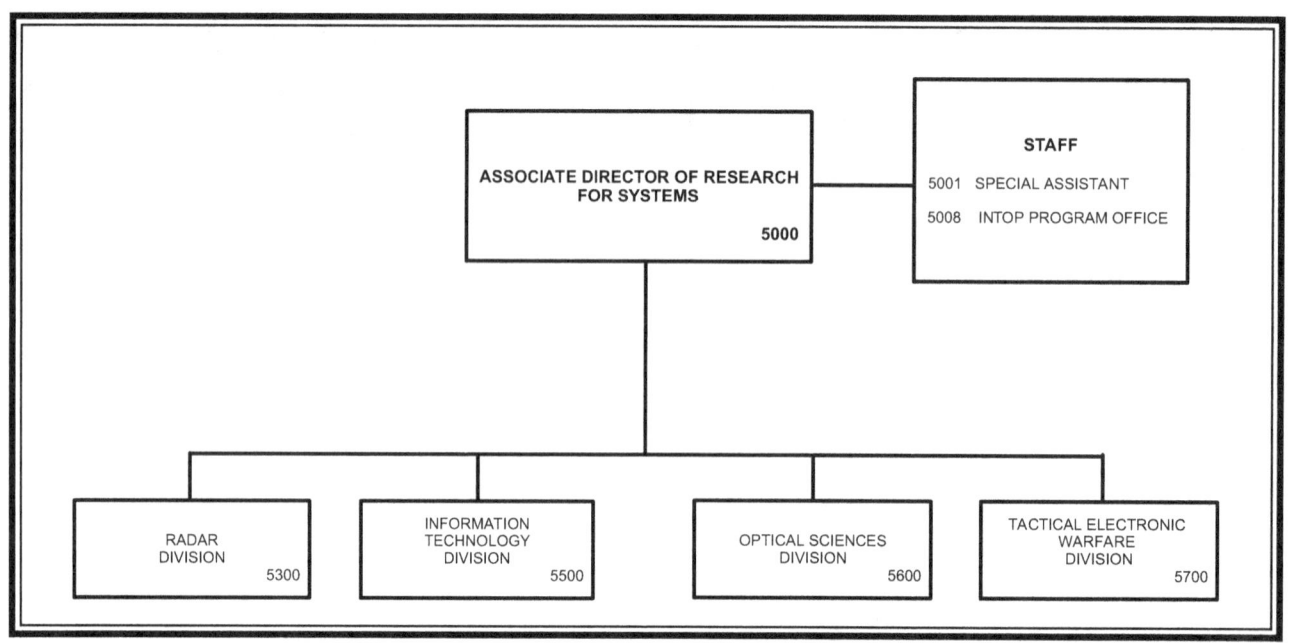

Key Personnel

Title	Code
Associate Director of Research for Systems	5000
Special Assistant	5001
Special Consultant	5007
Head, InTop Program Office	5008
Superintendent, Radar Division	5300
Superintendent, Information Technology Division	5500
Superintendent, Optical Sciences Division	5600
Superintendent, Tactical Electronic Warfare Division	5700

Point of contact: Code 5000A, (202) 767-3324

Radar Division

Code 5300
Staff Activity Areas

AEGIS coordination
Marine Corps/Air Force coordination

Maritime Domain Awareness
Multifunction RF systems

High-power millimeter-wave radar

Research Activity Areas

Radar Analysis

Target signature prediction
Electromagnetics and antennas
Airborne early-warning radar (AEW)
Inverse synthetic aperture radar (ISAR)
Sea clutter modeling
Periscope detection
Wideband array simulation and fabrication

Advanced Radar Systems

High-frequency over-the-horizon radar
Signal analysis
Real-time signal processing and equipment
Computer-aided engineering (CAE)
Array architecture optimization
FPGA-based digital processing
Future identification technology

Surveillance Technology

Shipboard surveillance radar
Ship self-defense
Electronic counter-countermeasures and
    electronic protection (EP)
Target signature recognition
Digital T/R modules
Asymmetric and expeditionary warfare
    spectrum management
Ultrawideband technology
Dynamic waveform diversity
Multistatic radar network
Information extraction
Ballistic missile defense
Mine detection

The Advanced Multifunction RF Concept (AMRFC) test bed is a proof-of-principle demonstration system capable of simultaneously transmitting and receiving multiple beams from common transmit and receive array antennas for radar, electronic warfare, and communications.

Wavelength scaled array: an ultrawideband array concept providing constant beamwidth across 8:1 bandwidth; designed by NRL-developed Domain Decomposition Algorithm.

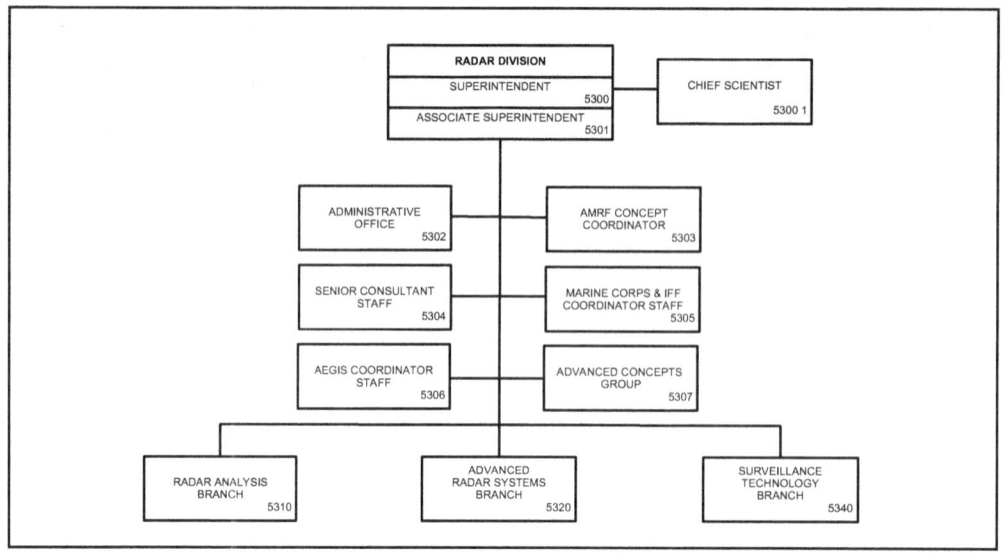

Basic Responsibilities

The Radar Division conducts research on basic physical phenomena of importance to radar and related sensors, investigates new engineering techniques applicable to radar, demonstrates the feasibility of new radar concepts and systems, performs related systems analyses and evaluation of radar, and provides special consultative services. The emphasis is on new and advanced concepts and technology in radar and related sensors that are applicable to enhancing the Navy's ability to fulfill its mission.

Personnel: 94 full-time civilian

Key Personnel

Title	Code
Superintendent, Radar Division	5300
Associate Superintendent	5301
Administrative Officer	5302
Senior Consultant Staff	5304
Marine Corps and IFF Coordinator	5305
AEGIS Coordinator	5306
Head, Advanced Concepts Group	5307
Head, Radar Analysis Branch	5310
Head, Advanced Radar Systems Branch	5320
Head, Surveillance Technology Branch	5340

Point of contact: Code 5300, (202) 404-2700

Information Technology Division

Code 5500
Research Activity Areas

Freespace Photonics Communications Office
Extended spectrum communications
Atmospheric channel effects on photonic transfer
Studies in marine miraging
Analog modulation techniques on freespace optical carriers
Modulating retroreflector based communications
Signature studies for ISR
Adaptive optics for freespace optical communications

Adversarial Modeling and Exploitation Office
Hostile intent and deception detection
Behavior detection research
Geospatial modeling and simulation
Dynamic semantic networks
Behavioral modeling, analysis and metrics
Spatially integrated social science
Integrated intelligence, surveillance, and reconnaissance
Automated video analysis and retrieval

Navy Center for Applied Research in Artificial Intelligence
Intelligent decision aids
Natural language and multimodal interfaces
Intelligent software agents
Machine learning and adaptive systems
Robotics software and computer vision
Neural networks
Novel devices/techniques for HCI
Spatial audio
Immersive simulation
Autonomous and intelligent systems
Case-based reasoning and problem-solving methods
Machine translation technology evaluation
Cognitive architectures
Human-robot interaction

Transmission Technology
Communication system architecture
Communication antenna/propagation technology
Communications intercept systems
Virtual engineering
Secure voice technology
Satellite and tactical networking
Satellite communications research
Satellite architecture analysis
RF systems analysis

Center for High Assurance Computer Systems
Secure service oriented architectures (SOA) and Secure Enterprise Architectures (SEA)
Formal specification/verification of system security
COMSEC application technology
Technology and solutions to secure networks and databases
Software engineering for secure systems
Key management and distribution solutions
Information systems security (INFOSEC) engineering
Formal methods for requirements specification and verification
Security product development
Secure wireless network and wireless sensor technology
Network security protocol modeling, simulation, and verificaton
Cross-domain solution technology development
Computer Network Defense (CND) technology

Hardware/software co-design
Malicious code analysis
Information hiding (watermarking, covert channel analysis, etc.)
Anonymizing systems
Quantum information science
Logical foundations of security

Networks and Communication Systems
Communication system engineering
Mobile, wireless networking technology
Bandwidth management (quality of service)
Joint service tactical networking
Integration of communication and C2 applications
Automated testing of highly mobile tactical networks
Reliable multicast protocols and applications
Communication network simulation
Networking protocols for directional antennas
Policy-based network management
Tactical voice-over IP
Sensor networks
Advanced tactical data links
Cognitive radio technology

Information Management and Decision Architectures
Virtual reality/mobile augmented reality
Visual analytics
Scientific visualization
Computer graphics
Human-computer interaction
Service oriented architecture
Service orchestration
Data and information management
Human-centered design
Parallel and distributed computation
Distributed modeling and simulation
Natural environments for distributed simulation
Intelligent decision support
Information sharing
Semantic web technology
Data mining
Software agents for data fusion

Center for Computational Science
Transparent optical network research and design
Parallel computing
Scalable high performance computing and networking for Navy and DoD
Large data in distributed computing
Scientific visualization
High-performance file systems
High-definition video technology
NRL labwide computer network and related services
Labwide support for web, email, and other information services
ATDnet and leading-edge WAN research networks

Ruth H. Hooker Research Library
Desktop/workbench access to relevant scientific resources
NRL scientific digital archive (TORPEDO)
Authoritative database of NRL-produced publications (NRL Online Bibliography)
Comprehensive literature/citation/classified searches
Extensive collection of print and digital books, journals, and technical reports

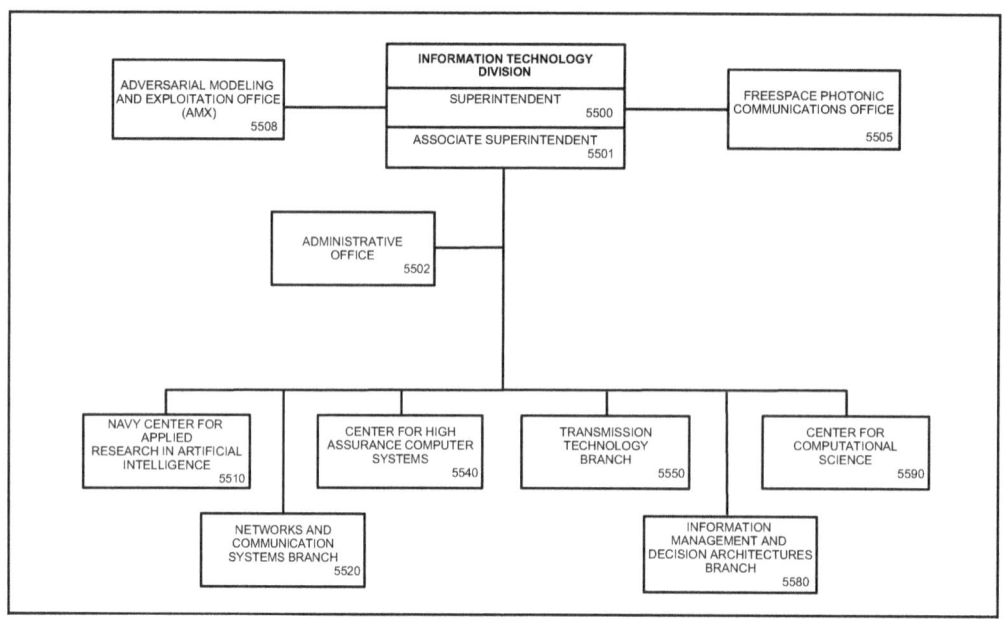

Basic Responsibilities

The Information Technology Division conducts basic research, exploratory development, and advanced technology demonstrations in the collection, transmission, processing, presentation, and distribution of information to provide information superiority and distributed networked force capabilities that improve Naval operations across all mission areas. The Division provides immediate solutions to current operational needs as required while developing those technologies necessary to implement the Navy after next.

Personnel: 204 full-time civilian

Key Personnel

Title	Code
Superintendent/NRL Chief Information Officer	5500
Associate Superintendent	5501
Administrative Officer	5502
Head, Freespace Photonic Communications Office	5505
Head, Adversarial Modeling and Exploitation Office	5508
Director, Navy Center for Applied Research in Artificial Intelligence	5510
Head, Networks and Communication Systems Branch	5520
Director, Center for High Assurance Computer Systems	5540
Head, Transmission Technology Branch	5550
Head, Information Management and Decision Architectures Branch	5580
Director, Center for Computational Science	5590
Chief Librarian, Ruth H. Hooker Research Library	5596

Point of contact: Code 5501, (202) 767-2954

Optical Sciences Division

Program analysis and development
Special systems analysis
Technical study groups

Technical contract monitoring
Theoretical studies

Research Activity Areas

Optical Materials and Devices
Advanced infrared optical materials
IR fiber-optic materials and devices
IR fiber chemical and environmental sensors
IR transmitting windows and domes
Transparent ceramic armor materials
Planar waveguide devices
IR nonlinear materials and devices
Ceramic laser gain materials
Advanced solar cell materials
Fiber lasers/sources and amplifiers
Radiation effects

Optical Physics
Laser materials diagnostics
Nonlinear frequency conversion
Optical instrumentation and probes
Optical interactions in semiconductor
  superlattices and organic solids
Laser-induced reactions
Organic light-emitting devices
Nanoscale electro-optical research
Aerosol optics

Applied Optics
UV, optical, and IR countermeasures
Ultraviolet component development
Missile warning sensor technology
UV, visible, and IR imager development
Multispectral/hyperspectral sensors
Multispectral/hyperspectral/detection algorithms
Framing reconnaissance sensors
Novel optical components
Sensor control and exploitation system
  development
IR low observables
EO/IR systems analysis
Atmospheric IR measurements
Airborne IR search and track technology

Photonics Technology
Fiber and solid-state laser/sources
High-speed (<100 fs) optical probing
High-power fiber amplifiers
High-speed fiber-optic communications
Antenna remoting
Free space communication
Photonic control of phased arrays
Micro-electro-optical-mechanical systems
Optical clocks
Microwave photonics

Optical Techniques
Fiber-optic materials and fabrication
Fiber Bragg grating sensors/systems
Fiber-optic sensors/systems (acoustic, magnetic,
  gyroscopes)
Integrated optics

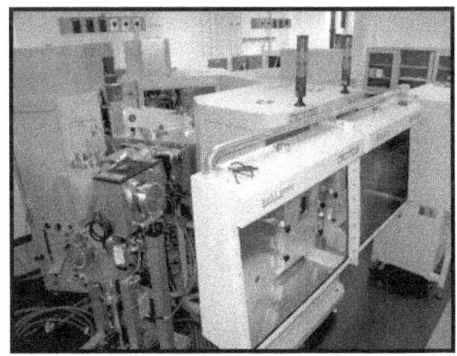

The Advanced Optical Materials Fabrication Laboratory, a state-of-the-art high vacuum cluster system, consists of a series of interconnected chambers allowing vacuum deposition of complex, multilayer films to be deposited and patterned without breaking vacuum during processing.

The Optical Fiber Preform Fabrication Facility includes computer control of the glass composition and standard fiber-optic dopants as well as rare earths, aluminum, and other components for specialty fibers.

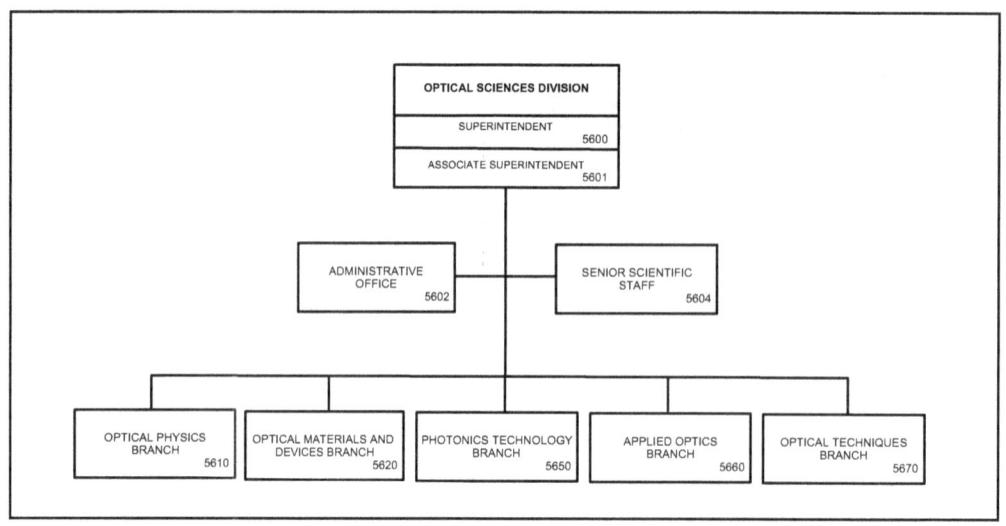

Basic Responsibilities

The Optical Sciences Division carries out a variety of research, development, and application-oriented activities in the generation, propagation, detection, and use of radiation in the wavelength region between near-ultraviolet and far-infrared wavelengths. The research, both theoretical and experimental, is concerned with discovering and understanding the basic physical principles and mechanisms involved in optical devices, materials, and phenomena. The development effort is aimed at extending this understanding in the direction of device engineering and advanced operational techniques. The applications activities include systems analysis, prototype system development, and exploitation of R&D results for the solution of optically related military problems. In addition to its internal program activities, the Division serves the Laboratory specifically and the Navy generally as a consulting body of experts in optical sciences. The work in the Division includes studies in quantum optics, laser physics, optical waveguide technologies, laser-matter interactions, atmospheric propagation, holography, optical data processing, fiber-optic sensor systems, optical systems, optical materials, radiation damage studies, IR surveillance and missile seeker technologies, IR signature measurements, and optical diagnostic techniques. A portion of the effort is devoted to developing, analyzing, and using special optical materials.

Personnel: 137 full-time civilian

Key Personnel

Title	Code
Superintendent, Optical Sciences Division	5600
Associate Superintendent	5601
Administrative Officer	5602
Head, Senior Scientific Staff	5604
Head, Optical Physics Branch	5610
Head, Optical Materials and Devices Branch	5620
Head, Photonics Technology Branch	5650
Head, Applied Optics Branch	5660
Senior Scientific Staff	5660.1
Head, Optical Techniques Branch	5670

Point of contact: Code 5602, (202) 767-6986

Tactical Electronic Warfare Division

Code 5700
Staff Activity Areas

EW Strategic Planning
Signature Technology Office

Effectiveness of Naval EW Systems (ENEWS)

Research Activity Areas

Offboard Countermeasures
Expendable technology and devices
Unmanned air vehicles
Offboard payloads
Decoys

Airborne Electronic Warfare Systems
Air systems development
Penetration aids
Power source development
Jamming and deception
Millimeter-wave technology
Communications CM

Ships Electronic Warfare Systems
Ships systems development
Jamming technology and deception
EW antennas
High power microwaves (HPM) research

Electronic Warfare Support Measures
Intercept systems and direction finders
RF signal simulators
Systems integration
Command and control interfaces
Signal processing

Advanced Techniques
Analysis and modeling simulation
Experimental systems
EW concepts
Infrared technology

Integrated EW Simulation
Hardware-in-the-loop simulation
Data management technology
Flyable ASM seeker simulators
Foreign materiel exploitation (FME)

EW Modeling and Simulation
High-fidelity threat models and simulations
Advanced system visualization
EW tactical decision aids
RF environmental and propagation modeling

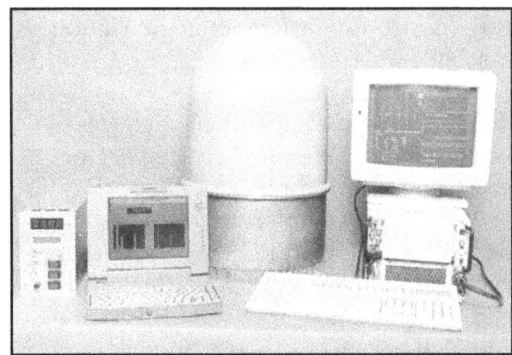

Using the latest composite, MMIC, and processing technologies, the Tactical Electronic Warfare Division has developed a small, lightweight, and inexpensive ESM receiving system for use on frigates, Coast Guard vessels, and various patrol aircraft.

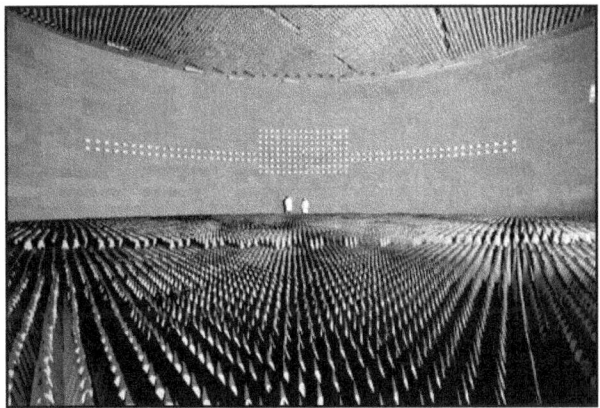

The Central Target Simulator (CTS) Programmable Array is part of a large hardware-in-the-loop simulation facility whose purpose is to test and evaluate electronic warfare systems and techniques used to counter radar-guided missile threats to Navy forces.

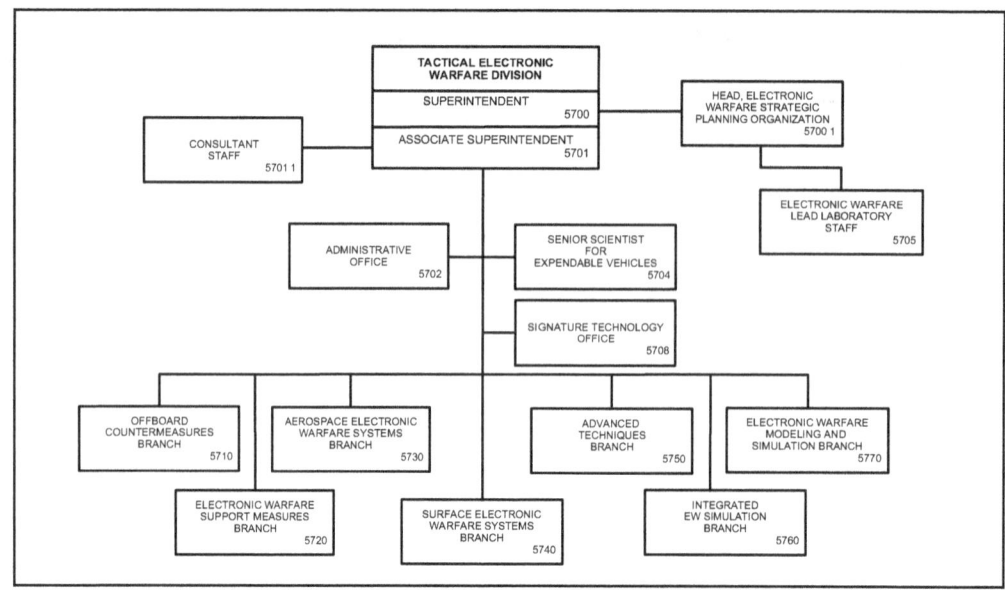

Basic Responsibilities

The Tactical Electronic Warfare Division (TEWD) is responsible for research and development in support of the Navy's tactical electronic warfare requirements and missions. These include electronic warfare support measures, electronic countermeasures, and supporting counter-countermeasures, as well as studies, analyses, and simulations for determining and improving the effectiveness of these systems.

Personnel: 237 full-time civilian

Key Personnel

Title	Code
Superintendent, Tactical Electronic Warfare Division	5700
Head, Electronic Warfare Strategic Planning Organization	5700.1
Associate Superintendent	5701
Administrative Officer	5702
Senior Scientist for Expendable Vehicles	5704
Head, Electronic Warfare Lead Laboratory Staff	5705
Head, Signature Technology Office	5708
Head, Offboard Countermeasures Branch	5710
Head, Electronic Warfare Support Measures Branch	5720
Head, Aerospace Electronic Warfare Systems Branch	5730
Head, Surface Electronic Warfare Systems Branch	5740
Head, Advanced Techniques Branch	5750
Head, Integrated Electronic Warfare Simulation Branch	5760
Head, Electronic Warfare Modeling and Simulation Branch	5770

Point of contact: Code 5701, (202) 767-5974

**Materials
Science and
Component
Technology
Directorate**

MATERIALS SCIENCE AND COMPONENT TECHNOLOGY DIRECTORATE

Code 6000

The Materials Science and Component Technology Directorate carries out a multidisciplinary research program whose objectives are the discovery, invention, and exploitation of new improved materials, the generation of new concepts associated with materials behavior, and the development of advanced components based on these new and improved materials and concepts. Theoretical and experimental research is carried out to determine the scientific origins of materials behavior and to develop procedures for modifying these materials to meet important naval needs for advanced platforms, electronics, sensors, and photonics.

The program includes investigations of a broad spectrum of materials including insulators, semiconductors, superconductors, metals and alloys, optical materials, polymers, plastics, artificially structured bio/molecular materials and composites, and energetic materials, which are used in important naval devices, components, and systems. New techniques are developed for producing, processing, and fabricating these materials for crucial naval applications.

The synthesis, processing, properties, and limits of performance of these new and improved materials in natural or radiation environments, and under deleterious conditions such as those associated with the marine environment, neutron or directed energy beam irradiation, or extreme temperatures and pressures, are established. For new materials design, emphasis is placed on protection of the environment.

Additionally, major thrusts are directed in advanced sensing, detection, reactive flow physics, computational physics, and plasma sciences. Areas of particular emphasis include nanoscience and technology, fluid mechanics and hydrodynamics, nuclear weapon effects simulations, high energy density materials including fuels, propellants, explosives, and storage devices, interactions of various types of radiation with matter, survivability of materials and components, and directed energy devices.

Dr. B.B. Rath was born in Banki, India. He received a B.S. degree in physics and mathematics from Utkal University, an M.S. in metallurgical engineering from Michigan Technological University, and a Ph.D. from the Illinois Institute of Technology.

Dr. Rath was Assistant Professor of Metallurgy and Materials Science at Washington State University from 1961 to 1965. From 1965 to 1972, he was with the staff of the Edgar C. Bain Laboratory for fundamental research of the U.S. Steel Corporation. From 1972 to 1976, he headed the Metal Physics Research Group of the McDonnell Douglas Research Laboratories in St. Louis, Missouri, until he came to NRL as Head of the Physical Metallurgy Branch. During this period, he was adjunct professor at Carnegie-Mellon University, the University of Maryland, and the Colorado School of Mines. Dr. Rath served as Superintendent of the Materials Science and Technology Division from 1982 to 1986, when he was appointed to his present position.

Dr. Rath is recognized in the fields of solid-state transformations, grain boundary migrations, and structure-property relationships in metallic systems. He has published over 140 papers in these fields and edited several books and conference proceedings.

Dr. Rath serves on several planning, review, and advisory boards for both the Navy and the Department of Defense, as well as for the National Materials Advisory Board of the National Academy of Sciences, National Science Foundation, University of Virginia, Colorado School of Mines, and the University of Florida. He is currently the Navy representative to the DOE Deputy Assistant Secretary's advisory and planning committee on methane hydrates, and the Navy representative to the Indo-U.S. Joint Commission on Science and Technology. He previously served as the Navy representative to the panel of The Technical Cooperation Program (TTCP) countries.

Dr. Rath is a member of the National Academy of Engineering. He is a fellow of the Minerals, Metals and Materials Society (TMS), American Society for Materials-International (ASM), Washington Academy of Sciences, Materials Research Society of India, the Institute of Materials of the United Kingdom, and the American Association for the Advancement of Science (AAAS). In 2007, Dr. Rath received an honorary doctorate in engineering from the Michigan Technological University and was elected to deliver the commencement address to the 2007 graduating class. In 2008, he received the Illinois Institute of Technology Mechanical Materials & Aerospace Engineering Department 2008 Alumni Recognition Award. In 2010, he received an honorary doctorate from Ravenshaw University.

Dr. Rath has received a number of honors and awards, most recently the Michigan Technological University Distinguished Alumni Award, the Padma Bhushan Award of Honors and Excellence bestowed by the President of India, and the Acta Materialia J. Herbert Hollomon Award. His other awards include the DoD Distinguished Civilian Service Award which is presented by the Secretary of Defense for distinguished accomplishments and sustained superior service, the 2005 Fred Saalfeld Award for Outstanding Lifetime Achievement in Science, the Presidential Rank Award for Distinguished Executive (2005), the NRL Lifetime Achievement Award (2004), National Materials Advancement Award from the Federation of Materials Societies (2001), the Presidential Rank of Meritorious Executive Award (1999 and 2004), the S. Chandrasekhar Award and Medal, and the Award of Merit for Group Achievement from the Chief of Naval Research. He received the 1991 George Kimball Burgess Memorial Award, the Charles S. Barrett Medal, and the prestigious TMS Leadership Award for his contributions to materials research. The American Society for Materials-International and The Metals, Minerals, and Materials Society have jointly recognized him with the TMS/ASM Joint Distinguished Lectureship in Materials & Society Award and the 2001 ASM Distinguished Life Membership Award. He has served as the 2004–2005 President of the American Society for Materials. He also has served as a member of the Boards of Directors/Trustees of TMS, ASM-International, and the Federation of Materials Society (FMS), as a member of the editorial boards of three international materials research journals, and as chairman of several committees of TMS, ASM, FMS, and American Association of Engineering Societies.

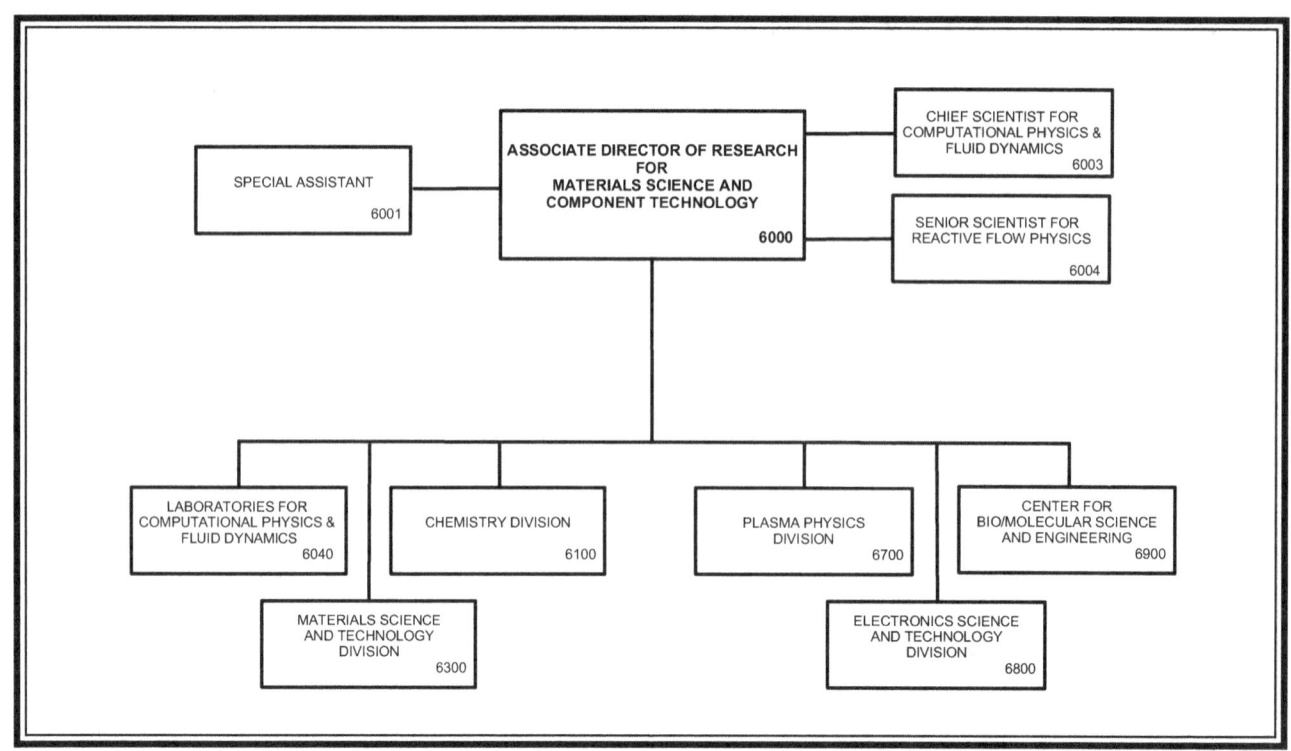

Key Personnel

Title	Code
Associate Director of Research for Materials Science and Component Technology	6000
Special Assistant	6001
Chief Scientist for Computational Physics and Fluid Dynamics	6003
Senior Scientist for Reactive Flow Physics	6004
Director, Laboratories for Computational Physics and Fluid Dynamics	6040
Superintendent, Chemistry Division	6100
Superintendent, Materials Science and Technology Division	6300
Superintendent, Plasma Physics Division	6700
Superintendent, Electronics Science and Technology Division	6800
Director, Center for Bio/Molecular Science and Engineering	6900

Point of contact: Code 6000, (202) 767-2538

Laboratories for Computational Physics and Fluid Dynamics

Code 6040
Research Activity Areas

Reactive Flows

Fluid dynamics in combustion
Turbulence in compressible flows
Multiphase flows
Turbulent jets and wakes
Turbulence modeling
Computational hydrodynamics
Propulsion systems analysis
Contaminant transport modelling
Fire and explosion mitigation

Computational Physics Developments

Laser-plasma interactions
Inertial confinement fusion
Solar physics modeling
Dynamical gridding algorithms
Advanced graphical and parallel
    processing systems
Electromagnetic and acoustic scattering
Microfluidics
Fluid structure interaction
Shock and blast containment

Olive (32P) and Snuffy (24P) — Origins at work.

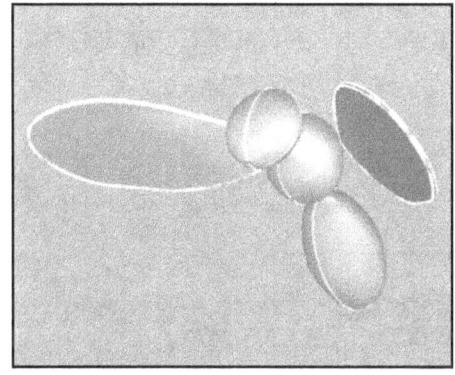

Unstructured grid technology has been used to obtain the surface pressure distribution on a hovering fruit fly *Drosophila*. Such computations are being carried out to gain insights into unsteady force production in nature that may guide in the design of insect-like autonomous air vehicles for the Navy.

This figure shows a contaminant cloud from a FAST3D-CT simulation of downtown Chicago using a 360 × 360 × 55 grid (6 m resolution). A 3 m/s wind off the lake from the left blows contaminant across a portion of the detailed urban geometry. The contaminant is lofted rapidly above the tops of the majority of the buildings due to their geometrical effect.

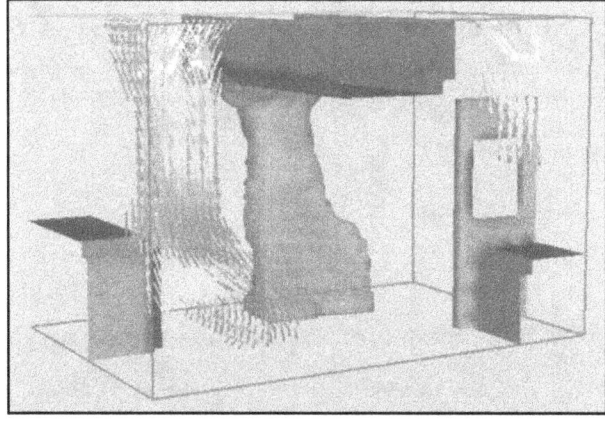

Water-mist trajectories and temperature distributions during the suppression of a fire inside a complex ship compartment. Simulations and experiments have shown that using fine water-mist can significantly reduce the amount of water needed for fire suppression.

Code 6040

Basic Responsibilities

The Laboratories for Computational Physics and Fluid Dynamics (LCP&FD) are responsible for the research leading to and the application of advanced analytical and numerical capabilities that are relevant to NRL, Navy, DoD, and other Government agencies. This research is pursued in the fields of compressible and incompressible fluid dynamics, reactive flows, fluid/structure interactions including submarine and aerospace applications, atmospheric and solar geophysics, magnetoplasma dynamics, application of parallel processing to large-scale problems such as unsteady flows of contaminants in and around cities, advanced propulsion concepts, flame dynamics for shipboard fire safety, jet noise reduction, and other disciplines of continuum computational physics as required to further the overall mission of NRL. The specific objectives of the LCP&FD are to develop and maintain state-of-the-art analytical and computational capabilities in fluid dynamics and related fields of physics; to establish in-house expertise in parallel processing for large-scale scientific computing; to perform analyses and computational experiments on specific relevant problems using these capabilities; and to transfer this technology to new and ongoing projects through cooperative programs with the research Divisions at NRL and elsewhere.

Personnel: 22 full-time civilian

Key Personnel

Title	Code
Director, Laboratories for Computational Physics and Fluid Dynamics	6040
Administrative Officer	6040.2
Chief Scientist for Computational Physics and Fluid Dynamics	6003
Senior Scientist for Reactive Flow Physics	6004
Head, Laboratory for Propulsion, Energetic, and Dynamic Systems	6041
Head, Laboratory for Advanced Computational Physics	6042
Head, Laboratory for Multiscale Reactive Flow Physics	6043

Point of contact: Code 6040.2, (202) 767-6581

Code 6100
Research Activity Areas

Chemical Diagnostics
Optical diagnostics of chemical reactions
Kinetics of gas phase reactions
Trace analysis
Atmosphere analysis and control
Ion/molecule processes
Environmental chemistry/microbiology
Methane hydrates
Laboratory on a chip
Alternate energy sources

Materials Chemistry
Synthesis and evaluation of
  innovative polymers and composites
Functional organic coatings
Polymer characterization
Magnetic resonance
Degradation and stabilization mechanisms
High-temperature resins
Bio-inspired materials
Novel nanotubes and nanofibers
Reactive nanometals

Center for Corrosion Science and Engineering
Materials failure analysis
Marine coatings
Cathodic protection

Corrosion science
Environmental fracture and fatigue
Corrosion control engineering

Surface/Interface Chemistry
Tribology
Surface properties of materials
Surface/interface analysis
Chemical/biological sensors
Surface reaction dynamics
Adhesion
Bio/organic interfaces
Diamond films
Energy storage materials
Nanostructured materials and interfaces
Electrochemistry
Plasmonics
Synchrotron radiation applications

Safety and Survivability
Combustion dynamics
Fire protection and suppression
Personnel protection
Modeling and scaling of combustion systems
Mobility fuels
Chemometrics/data fusion
Trace analysis

The Key West site of the NRL Center for Corrosion Science and Engineering specializes in understanding and modeling the marine environment's impact on naval materials. A complete laboratory for the study of corrosion control technologies provides sponsors with prototypical seawater exposure of their systems.

The ex-USS *Shadwell* (LSD 15), moored in Mobile Bay, Alabama, is NRL's full-scale, advanced fire research vessel operated by the Chemistry Division.

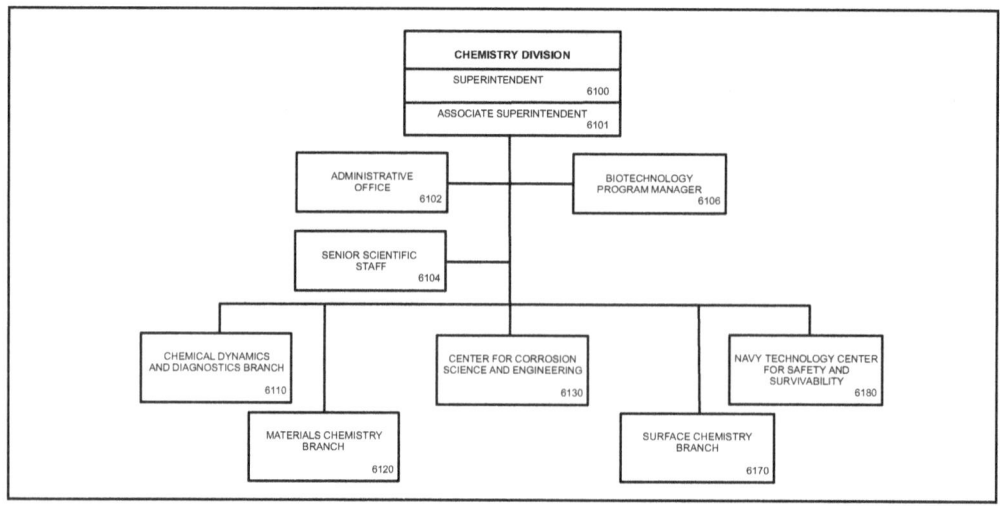

Basic Responsibilities

The Chemistry Division conducts basic research, applied research, and development studies in the broad fields of chemical/structural diagnostics, reaction rate control, materials chemistry, surface and interface chemistry, corrosion passivation, environmental chemistry, and ship safety/survivability. Specialized programs within these fields include coatings, functional polymers/elastomers, clusters, controlled release of energy, physical and chemical characterization of surfaces, electrochemistry, assembly and properties of nanometer structures, tribology, chemical vapor deposition/etching, atmosphere analysis and control, environmental protection/reclamation, prevention/control of fires, mobility fuels, modeling/simulation, and miniaturized sensors for chemical, biological, trace analysis and data fusion, and explosives.

To enhance protection of Navy personnel and platforms from damage and injury in peace and wartime, the Navy Technology Center for Safety and Survivability performs RDT&E on fire and personnel protection, fuels, chemical defense, submarine atmospheres, and damage control aspects of ship and aircraft survivability; supports Navy and Marine Corps requirements in these areas; and acts as a focus for technology transfer in safety and survivability.

To address problems in corrosion and marine fouling, a Marine Corrosion Facility is located in Key West, Florida. This laboratory resides in an unparalleled site for natural seawater exposure testing and marine related materials evaluation. The tropical climate is ideal for marine exposure testing. Along with the high quality seawater, the location provides small climatic variation and a stable biomass throughout the year.

Personnel: 111 full-time civilian; 2 military; 6 intermittent; 2 part-time

Key Personnel

Title	Code
Superintendent, Chemistry Division	6100
Associate Superintendent	6101
Administrative Officer	6102
Senior Scientific Staff	6104
Senior Scientific Staff	6104
Biotechnology Program Manager	6106
Head, Chemical Dynamics and Diagnostics Branch	6110
Head, Materials Chemistry Branch	6120
Head, Center for Corrosion Science and Engineering	6130
Head, Surface Chemistry Branch	6170
Head, Navy Technology Center for Safety and Survivability	6180

Point of contact: Code 6102, (202) 767-2460

Materials Science and Technology Division

Code 6300
Research Activity Areas

Spintronics
Materials and Sensors
Superconducting materials
Magnetic materials
Optoelectronic materials
Electroceramic materials
Radar absorbing materials
THz sources and detectors
Bioelectronics
Remote video surveillance
Chemical sensors
Chaos theory
Thin film deposition
  Pulsed laser deposition
  Ion-beam-assisted deposition
  Variable balance magnetron sputtering
Laser direct write
Ion implantation
Glass fiber draw tower
Polymer synthesis and characterization
Precision calorimetry
Analysis of extrasolar materials
Ballistic materials
Personal protective equipment
Explosives detection

Multifunctional Materials
Biomechanical surrogate development for
  threat response characterization
Biomechanical simulation
Composite material systems
  Multifunctional structure + other (e.g., power, etc.)
  Hierarchical and tiled architectures
  Armor protection

Corrosion simulation and control
  Modeling of electrochemical corrosion systems
  Evaluation of cathodic protection performance
Image-based modeling
Materials by design
Mesoscale material characterization and
  simulation
Physical metallurgy
  Ferrous, nonferrous, and intermetallic alloys
  Hot/cold isostatic pressing
  Micro/nanostructure characterization
  Three-dimensional microstructure characterization
  Synthesis/processing of metal
  Rapid solidification
  Welding/joining technology
  Heat treating and phase transformations
Synthesis and processing of advanced ceramics
  High energy density dielectrics
  Piezoelectrics

Computational Materials
Condensed matter theory
Electronic structure of solids and clusters
Molecular dynamics
Quantum many-body theory
Theory of magnetic materials
Theory of alloys
Semiconductor and surface physics
Theoretical studies of phase transitions
Atomic physics theory
Protein modeling
Continuum multiphysics modeling

The variable pressure scanning electron microscope facility provides capabilities for imaging down to 10 nm resolution, with both secondary and backscattered electron detectors. The capability of operating at variable pressures allows for the examination of nonconducting samples without the need for coating. The system is equipped with energy dispersive spectroscopy (EDS) capabilities for measuring, quantifying, and mapping chemical composition, as well as an electron backscattered diffraction (EBSD) camera for the mapping and quantification of material crystallography.

Five-axis laser micromachining and laser direct-write system based on a high-repetition-rate (100 kHz) UV solid-state laser (266 nm). This system can directly deposit and pattern metals and dielectrics on doubly curved surfaces (such as the hemispherical dome shown) with a linewidth resolution down to a few microns and a positional accuracy of one micron.

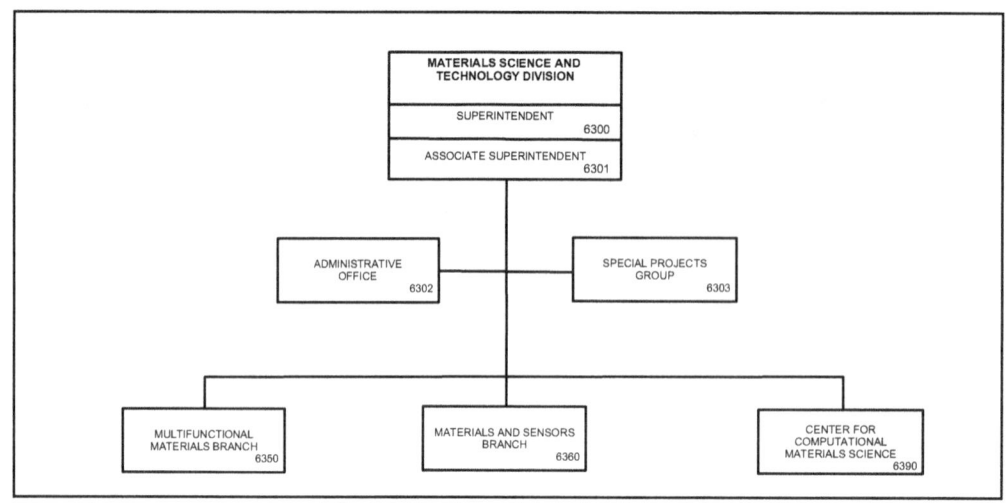

Basic Responsibilities

The Materials Science and Technology Division conducts basic and applied research and engages in exploratory and advanced development of materials having substantive value to the Navy. R&D programs encompass the intrinsic behavior of metals, insulators, composites, and ceramics, including efforts in ferrous alloys, intermetallic compounds, superconducting, dielectric, and magnetic materials, films and coatings, and multifunctional materials systems. The programs encompass advanced synthesis and processing techniques as well as postprocessing techniques to fabricate sensors, devices, structures, and components. A variety of state-of-the-art characterization tools are used to probe the atomic and microstructure nature (composition and structure) of the materials as well as to delineate the fundamental properties of the material or material system. Response of materials and material systems to a variety of external influences (mechanical, chemical, optical, electromagnetic radiation, high-power lasers, temperature, etc.) is integral to the Division's programs, as are performance and reliability projections for military service lifetime. The program includes strong theoretical, computational, and simulation efforts to predict, guide, and explain the behavior of materials and materials systems. Studies conducted in the Division provide guidance for the selection, design, certification, and life-cycle management of material in naval vehicles and systems. The diversity of R&D programs in the Division is carried out by multidisciplinary teams of materials scientists, metallurgists, ceramists, physicists, chemists, and engineers using the most advanced testing facilities and diagnostic techniques.

Personnel: 110 full-time civilian

Key Personnel

Title	Code
Superintendent, Materials Science and Technology Division	6300
Senior Scientist	6300.1
Associate Superintendent	6301
Administrative Officer	6302
Head, Special Projects Group	6303
Head, Multifunctional Materials Branch	6350
Head, Materials and Sensors Branch	6360
Head, Center for Computational Materials Science	6390

Point of contact: Code 6302, (202) 767-2458

Plasma Physics Division

Code 6700
Research Activity Areas

Radiation Hydrodynamics
Radiation hydrodynamics of Z-pinches and laser-produced plasmas
X-ray source development
Cluster dynamics in intense laser fields
X-ray channeling and propagation
Plasma kinetics for directed energy and fusion
Plasma discharge physics
Dense plasma atomic physics, equation of state
Numerical simulation of high-density plasma
Laser driven ion/neutron sources

Laser Plasma
Nuclear weapons stockpile stewardship
Laser fusion, inertial confinement
Megabar high-pressure physics
Rep-rate KrF laser development
Impact fusion
Laser fusion technology
Laser fusion energy
Detection of chemical/biological/nuclear materials

Charged Particle Physics
Applications of modulated electron beams
Rocket, satellite, and shuttle-borne natural and active experiments
Laboratory simulation of space plasma processes

Large-area plasma processing sources
Plasma processing of energy sensitive materials
Atmospheric and ionospheric GPS sensing
Ionospheric effects on communications
Electromagnetic launchers
Radiation belt remediation

Pulsed Power Physics
Production, focusing, and propagation of intense electron and ion beams
High-power, pulsed radiography
Plasma radiator and bremsstrahlung diode sources
Capacitive and inductive energy storage
Nuclear weapons effects simulation
Electromagnetic launchers
Detection of Special Nuclear Materials
Advanced energetics via stimulated nuclear decay

Beam Physics
Advanced accelerators and radiation sources
Microwave, plasma, and laser processing of materials
Microwave sources: magnicons and gyrotrons
Nonlinear dynamics of coupled lasers
Ultrahigh-intensity laser-matter interactions
Free electron lasers and laser synchrotrons
Theory and simulation of space and solar plasmas
Global ionospheric and space weather modeling
Laser propagation in the atmosphere
Underwater laser interactions

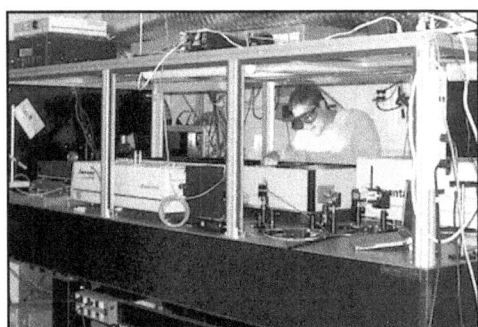

The NRL Ti:Sapphire Femtosecond Laser (TFL) currently operates at 50 fsec, 10 TW and provides a facility to conduct research in intense laser-plasma interactions, ultrashort intense laser propagation in the atmosphere, remote sensing of chem/bio agents, and laser-induced electrical discharges.

Nike is the world's largest krypton fluoride (KrF) laser and is used to explore physics issues for laser fusion. Shown is the propagation bay where 56 short-duration (4–5 ns) beams are directed by mirrors first to the electron-beam-pumped amplifiers and then to the target facility. The Nike KrF system achieves extremely uniform high-intensity illumination of planar targets by overlapping numerous smoothed laser beams. Typical experiments include studies of the ablative acceleration of matter to high velocities (100 km/sec) and studies of the reaction of materials to very high pressures (10 million atmospheres) produced by the laser light.

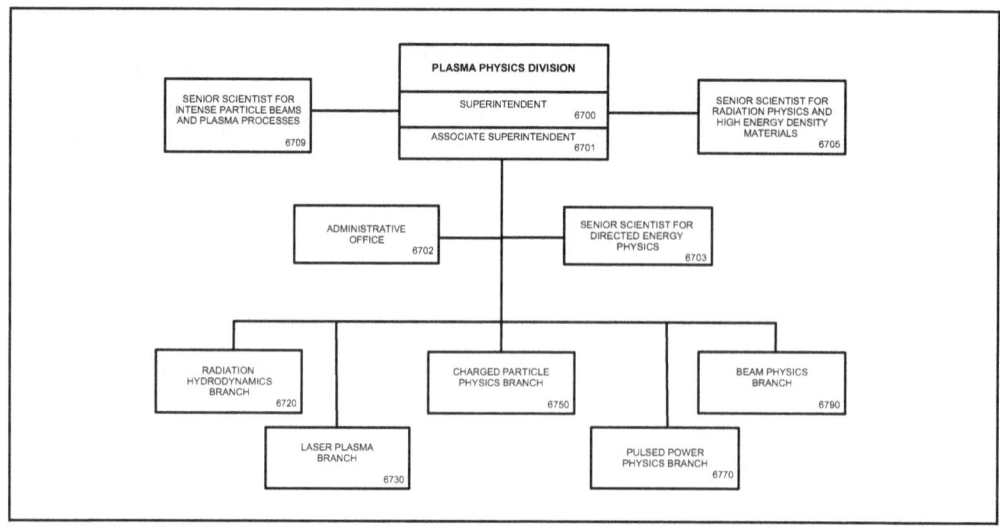

Basic Responsibilities

The Plasma Physics Division conducts a broad theoretical and experimental program of basic and applied research in plasma physics, laboratory discharge, and space plasmas, intense electron and ion beams and photon sources, atomic physics, pulsed power sources, laser physics, advanced spectral diagnostics, and nonlinear systems. The effort of the Division is concentrated on a few closely coordinated theoretical and experimental programs. Considerable emphasis is placed on large-scale numerical simulations related to plasma dynamics; ionospheric, magnetospheric, and atmospheric dynamics; nuclear weapons effects; inertial confinement fusion; atomic physics; plasma processing; nonlinear dynamics and chaos; free electron lasers and other advanced radiation sources; advanced accelerator concepts; and atmospheric laser propagation. Areas of experimental interest include laser-plasma, laser-electron beam, and laser-matter interactions, high-energy laser weapons, laser shock hydrodynamics, thermonuclear fusion, electromagnetic wave generation, the generation of intense electron and ion beams, large-area plasma processing sources, electromagnetic launchers, high-frequency microwave processing of ceramic and metallic materials, advanced accelerator development, inductive energy storage, laboratory simulation of space plasma phenomena, high-altitude chemical releases, and in situ and remote sensing space plasma measurements.

Personnel: 85 full-time civilian

Key Personnel

Title	Code
Superintendent, Plasma Physics Division	6700
Associate Superintendent	6701
Administrative Officer	6702
Senior Scientist, Directed Energy Physics	6703
Senior Scientist, Radiation Physics and High Energy Density Materials	6705
Senior Scientist, Intense Particle Beams and Plasma Processes	6709
Head, Radiation Hydrodynamics Branch	6720
Head, Laser Plasma Branch	6730
Head, Charged Particle Physics Branch	6750
Head, Pulsed Power Physics Branch	6770
Head, Beam Physics Branch	6790

Point of contact: Code 6700, (202) 767-2723

Electronics Science and Technology Division

Code 6800
Research Activity Areas

Electronic Materials
Preparation and development of magnetic, dielectric, optical, and semiconductor materials including micro- and nanostructures

Electrical, optical, and magneto-optical studies of semiconductor microstructures and nanostructures, superlattices, surfaces, and interfaces

Impurity and defect studies

Surface research and interface physics

Theoretical solid-state physics

Microwave Technology
Microwave and millimeter-wave integrated circuits and components research

High-frequency device design, simulation, and fabrication

Reliability and failure physics of electronic devices and circuits

Oxide- and carbon-based electronics for high-frequency devices

Power Electronics
Power device design, simulation, and fabrication

High-voltage/high-temperature power device and components research

Growth and characterization of wide bandgap and thin film materials for power devices

Wafer bonding for power devices and novel substrates

Reliability and failure physics of power devices

Nanoelectronics
Characterization of nanosurfaces and interfaces

Nanoelectronic device research and fabrication

Processing research for nanometric devices

Radiation Effects
Space experiments and satellite survivability

Single event and total ionizing dose effects

Radiation hardening of electronics devices, circuits, and optoelectronic sensors

Ultrafast charge collection

Environmental hazard remediation

Advanced photovoltaic technologies

Femtosecond laser research

Radiation effects in microelectronics and photonics

Solid-State Devices
Solid-state optical sensors

Photovoltaic research and development

Mid- and far-infrared photodiodes/arrays

Microelectronics device research and fabrication

Solid-state circuits research

Signal processing research

Vacuum Electronics
Compact millimeter-wave power amplifier research and development

Cathode research and electron emission science

Materials development for microwave and millimeter-wave applications

Development of microfabrication techniques for upper millimeter-wave devices

Theory and numerical techniques for modeling of fast-wave and slow-wave devices

Techniques for broadband, complex waveform generation and analysis for high data rate communications and electronic warfare

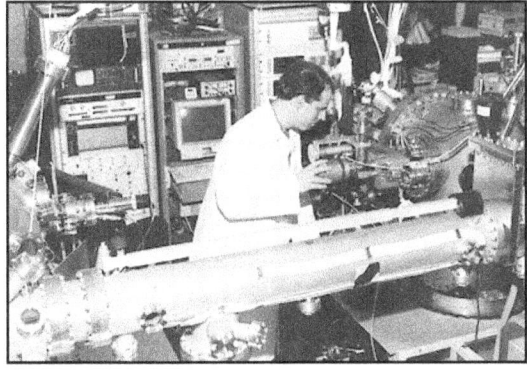

The EPICENTER specializes in molecular beam epitaxial growth of nanostructures created by alternating layers of narrow bandgap materials made available from four ultrahigh-vacuum chambers. These structures are expected to improve the performance of far-infrared detectors, midwave lasers, and superhigh frequency transistors and resonant tunneling diodes. Here a scientist creates a structure using high-vacuum, chamber-to-chamber sample transfer.

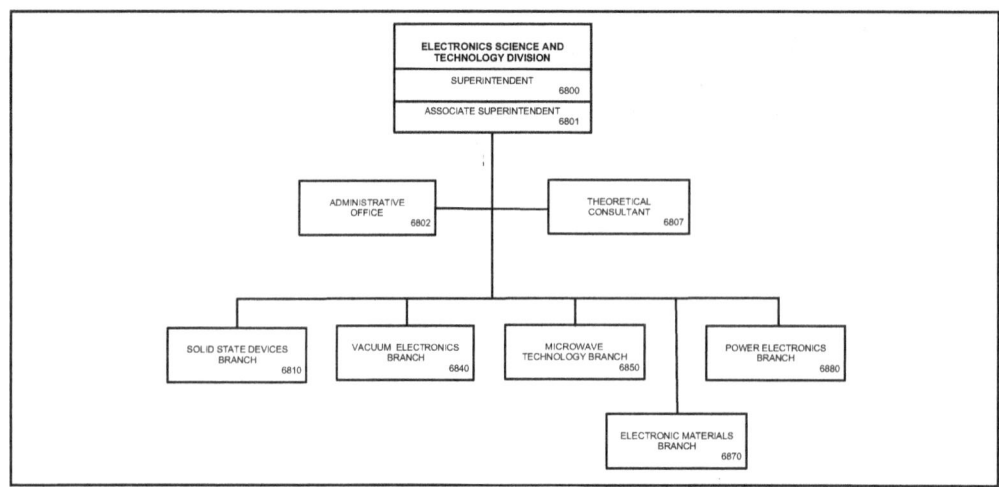

Basic Responsibilities

The Electronics Science and Technology Division conducts programs of basic science and applied research and development in materials growth and properties, surface physics, micro- and nanostructure electronics, microwave techniques, microelectronic device research and fabrication, vacuum electronics, and cryoelectronics, including superconductors. The activities of the Division integrate device research with basic materials investigations and with systems research and development needs.

Personnel: 98 full-time civilian

Key Personnel

Title	Code
Superintendent, Electronics Science and Technology Division	6800
Associate Superintendent	6801
Administrative Officer	6802
Senior Scientist for Nanoelectronics	6877
Head, Solid State Devices Branch	6810
Head, Vacuum Electronics Branch	6840
Head, Microwave Technology Branch	6850
Head, Electronic Materials Branch	6870
Head, Power Electronics Branch	6880

Point of contact: Code 6802, (202) 767-3416

Code 6900
Research Activity Areas

Biologically Derived Microstructures
Self-assembly, molecular machining
Synthetic membranes
Nanocomposites
Tailored electronic materials
Low observables
Molecular engineering, biomimetic materials
Molecular imprinting
Viral scaffolds
Multifunctional decontamination coatings

Biosensors
Binding polypeptides and proteins
Cell-based biosensors
DNA biosensors
Fiber-optic biosensors
Flow immunosensors
Array-based sensors
Optical biosensors
Microfluidics

Novel Materials
Soil/groundwater explosives detection
Antifouling paint, controlled release
Single chain antibodies
Liquid crystal nanoparticles
Liquid crystal elastomers
Nano- and mesoporous materials
Quantum dot and protein conjugates
Biomimetic materials

Molecular Biology
Genomics and proteomics of marine bacteria
Tissue engineering
Gene arrays, biomarkers
System and synthetic biology

Energy Harvesting
Biomaterials for charge storage
Ocean floor biofuel cell
Photo-induced electron transfer

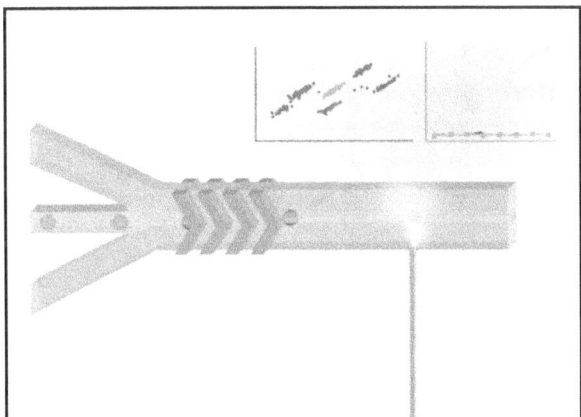

Microfluidic structures direct arrays of beads one-by-one past a laser beam. If a biothreat is bound to the surface of a bead, the identity of the threat can be determined by the color code on the bead.

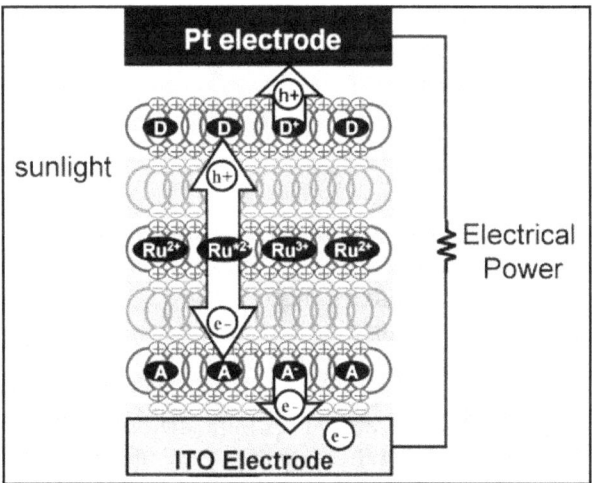

Utilizing the self-assembly of molecular chromophores, electron acceptors, and electron donors to investigate non-silicon-based methods for electricity generation from sunlight.

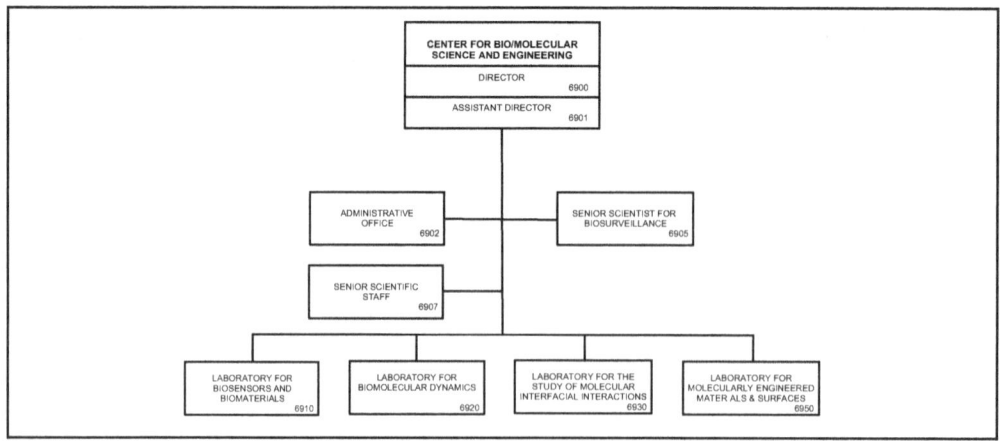

Basic Responsibilities

The Center for Bio/Molecular Science and Engineering is using the tools of modern biology, physics, chemistry, and engineering to develop advanced materials and sensors. The long-term research goal is first to gain a fundamental understanding of the relationship between molecular architecture and the function of materials, then apply this knowledge to solve problems for the Navy and DoD community. The key theme is the study of complex bio/molecular systems with the aim of understanding how "nature" has approached the solution of difficult structural and sensing problems. Technological areas currently being studied include molecular and microstructure design, molecular biology, self-assembly, controlled release and encapsulation, and surface patterning and modification. Much of the research deals with the self-assembly of lipids, proteins, and liquid crystals into complex microstructures for use in advanced material applications, and the harnessing of the recognition functions of proteins and cells for the development of advanced sensors. A highly multidisciplinary staff is required to pursue these research and development programs. The Center provides a stimulating environment for cross-disciplinary programs in the areas of immunology, biochemistry, electrochemistry, inorganic and polymer chemistry, microbiology, microlithography, photochemistry, biophysics, spectroscopy, advanced diagnostics, organic synthesis, and electro-optical engineering.

Personnel: 57 full-time civilian

Key Personnel

Title	Code
Director, Center for Bio/Molecular Science and Engineering	6900
Assistant Director	6901
Administrative Officer	6902
Senior Scientist for Biosurveillance	6905
Head, Senior Scientific Staff	6907
Head, Laboratory for Biosensors and Biomaterials	6910
Head, Laboratory for Biomolecular Dynamics	6920
Head, Laboratory for the Study of Molecular Interfacial Interactions	6930
Head, Laboratory for Molecularly Engineered Materials and Surfaces	6950

Point of contact: Code 6902, (202) 404-6012

Ocean and Atmospheric Science and Technology Directorate

OCEAN AND ATMOSPHERIC SCIENCE AND TECHNOLOGY DIRECTORATE

Code 7000

The Ocean and Atmospheric Science and Technology Directorate performs research and development in the fields of acoustics, remote sensing, oceanography, marine geosciences, marine meteorology, and space science. Areas of emphasis in acoustics include advanced acoustic concepts and computation, acoustic signal processing, physical acoustics, acoustic systems, ocean acoustics, and acoustic simulation and tactics. Areas of emphasis in remote sensing include radio, infrared, and optical sensors, remote sensing physics and hydrodynamics, remote sensing simulation, and imaging systems. Areas of emphasis in oceanography include coastal and open ocean dynamics, ocean modeling and prediction, coastal and open ocean processes, remote sensing applications to oceanography, and marine biocorrosion processes. Areas of emphasis in marine geosciences include marine physics, seafloor sciences, geospatial information science and technology, and mapping, charting, and geodesy. Areas of emphasis in marine meteorology include atmospheric dynamics for theater-wide, tactical-scale prediction systems and forecast support, and meterological applications development. Areas of emphasis in space science include middle and upper atmosphere physics, solar terrestrial relationships, solar physics, and higher energy astronomy. Senior naval officers are assigned as military advisors to help maintain the directorate focus on operational Navy and other DoD requirements in these areas of emphasis. The directorate is responsible for administrative and technical support to major activities in Washington, DC; Stennis Space Center, Mississippi; and Monterey, California.

Dr. E.R. Franchi was born in Huntington, New York. He graduated from Clarkson University in 1968 with a bachelor of science degree in mathematics. He received his master of science (1970) and Ph.D. (1973) degrees, both in applied mathematics, from Rensselaer Polytechnic Institute. After completing his graduate studies, Dr. Franchi accepted a research position with Bolt, Beranek, and Newman where he performed validation studies of underwater acoustic propagation and noise models.

Dr. Franchi joined the Naval Research Laboratory in 1975 as a research mathematician in the Acoustics Division. In this position, he conducted and directed research in low frequency acoustic reverberation and scattering, including design and conduct of field experiments, development of signal processing techniques, data analysis and interpretation, computer prediction models, and active sonar performance studies. In 1986, he was named Head of the Acoustic Systems Branch where he was responsible for programs that emphasized theoretical, experimental, and computational research to understand the physical mechanisms of acoustic propagation, scattering, and ambient noise that control the design and performance of large-aperture passive sonar systems, low frequency active sonar systems, and shallow water sonar systems.

In July 1988, Dr. Franchi was appointed to the Senior Executive Service and selected as the Associate Technical Director of the Naval Ocean Research and Development Activity (NORDA) and its Director of Ocean Acoustics and Technology. The Directorate conducted basic, exploratory, and advanced research and development and program management in the areas of acoustic model development and simulation, ocean acoustics measurements, and ocean engineering in support of all undersea warfare missions. In October 1992, the Directorate became the Center for Environmental Acoustics in the Acoustics Division of the Naval Research Laboratory, with Dr. Franchi as Director. Dr. Franchi was selected to the position of Superintendent of the Acoustics Division in October 1993. The Acoustics Division conducts basic, exploratory, and applied research and development in areas of acoustic modeling and simulation, ocean acoustics measurements, acoustic systems development, acoustic signal processing, and physical acoustics. He was responsible for the technical/ scientific management, direction, and administration of programs with a total budget in excess of $25M, and for efficient management of division resources including the activities of approximately 110 civilian personnel. He served as Acting Associate Director of Research for the Ocean and Atmospheric Science and Technology Directorate from October 2001 to May 2002 and from June 2007 to April 2008. In April 2008, he was selected as the Associate Director of Research.

Dr. Franchi received the Presidential Rank Award of Meritorious Executive in 2003. He has over 35 years experience in underwater acoustics research and is the author/co-author of over 35 publications. He is recognized as an authority on underwater acoustic scattering and reverberation and has played major roles in Navy low frequency active sonar programs as both performer and advisor/consultant. He served as the U.S. National Leader of The Technical Cooperation Program's multinational Panel on ASW Systems and Technology from 1996 to 2002, and served as its Panel Chairman from 2002 to 2009. In 2011, Dr. Franchi received the TTCP Personal Achievement Award in recognition of his significant contributions and strategic vision in leading the ASW Panel. He represents the United States to the NATO Undersea Research Centre Scientific Committee of National Representatives and served as its Committee Chairman from 2010 to the present. In 2011, he was appointed to the NATO Science and Technology Reform Implementation Team. He was elected to Pi Mu Epsilon, the Honorary National Mathematics Society, while an undergraduate at Clarkson University. Dr. Franchi is a member of the Acoustical Society of America and past member of the Mathematical Association of America. Since 2004, he has volunteered his time to serve on the Board of Directors of the NRL Federal Credit Union.

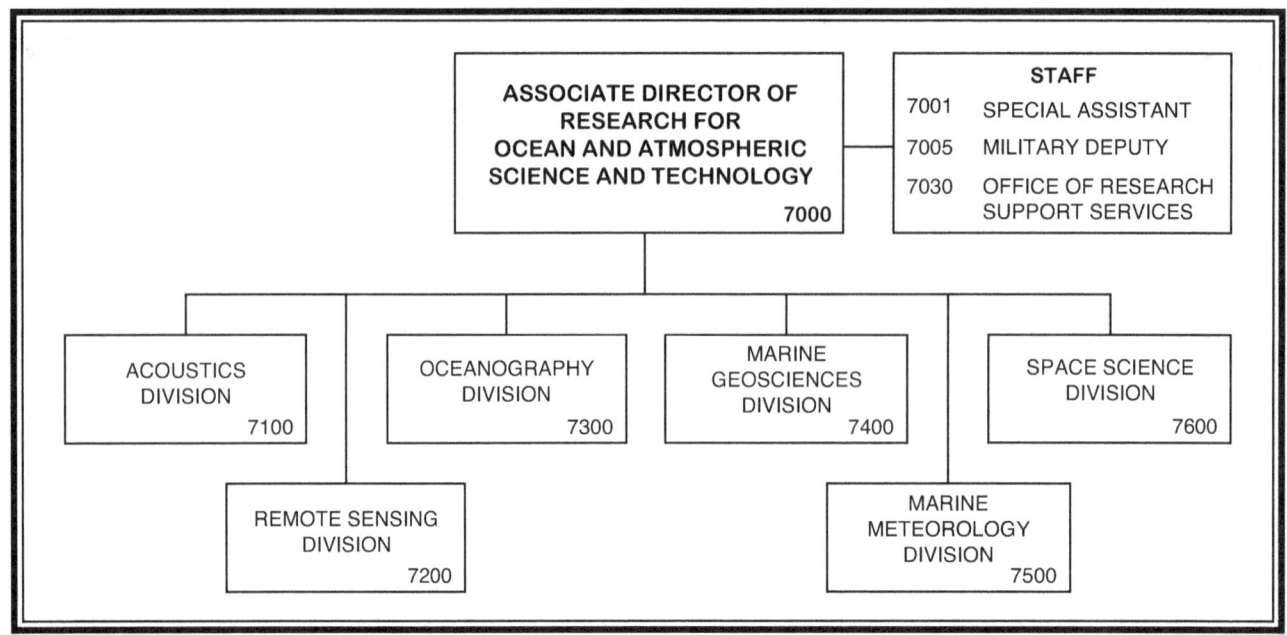

Key Personnel

Title	Code
Associate Director of Research for Ocean and Atmospheric Science and Technology	7000
Special Assistant	7001
Military Deputy	7005
Head, Office of Research Support Services	7030
Superintendent, Acoustics Division	7100
Superintendent, Remote Sensing Division	7200
Superintendent, Oceanography Division	7300
Superintendent, Marine Geosciences Division	7400
Superintendent, Marine Meteorology Division	7500
Superintendent, Space Science Division	7600

Point of contact: Code 7000A, (202) 404-8174

Office of Research Support Services (NRL-SSC)

Code 7030
Staff Activity Areas

Office of Research Support
Conference coordination, video teleconferencing
Directives, reports, forms

Facilities Office
Facilities planning and maintenance
Vehicles

HPC Management Office
Supercomputing interface management

Safety/Environmental Office
Industrial/laboratory safety
Specialized safety training
Hazard abatement
Mishap prevention
Hazardous materials program
Hazardous waste disposal

Public Affairs Office
Community relations
News releases
Exhibits
Information
Freedom of Information Act

NRL-SSC Network Management Office
Data communications
Data networking
Computer network maintenance

Basic Responsibilities

The Office of Research Support Services is responsible for the operational and management support necessary for the day-to-day operations at NRL Stennis Space Center, Mississippi (NRL-SSC). The Head of NRL-SSC acts for the Commanding Officer in dealing with local Navy, Federal, and civil activities and personnel on matters relating to NRL-SSC support activities and facilities, community and multicommand issues, and safety and disaster control measures.

Support functions include public affairs, network support, safety, high performance computer management, and support services to include management, administration, and facilities.

Personnel: 8 full-time civilian

Key Personnel

Title	Code
Head, Office of Research Support Services	7030
Administrative Officer	7030.2
Head, Facilities Office	7030.3
Public Affairs Officer	7030.4
Safety/Environmental Officer	7030.5
HPC Management Office	7030.6
NRL-SSC Network Management Office	7030.8

Point of contact: Code 7030, (228) 688-4010; DSN 828-4010

Acoustics Division

Code 7100
Research Activity Areas

Physical Acoustics
Structural acoustics
Quantum effects in phononic crystals
Nanomechanical devices
Fiber-optic acoustic sensors
Acoustic transduction
Inverse scattering
Target strength/radiation modeling
Flow-induced noise and vibration
Active sonar classification
Underwater distributed, networked sensing
AUV-based sensing

At-sea deployment of underwater acoustic communications source/receiver array. The purpose is to conduct multiple-input-multiple-output (MIMO) underwater acoustic communications experiments to increase the bandwidth for distributed systems.

Structural acoustic studies are conducted in the one-million-gallon Acoustic Holographic Pool Facility.

Acoustic Signal Processing and Systems
Underwater acoustic communications and
  networking
Limits of array performance
Waveguide invariant processing
Acoustic field uncertainty
Acoustic interactions with transonic/
  supersonic flows
Acoustic noise forecasting
Long-range underwater communications
Underwater distributed sensing networks
Ocean boundary scattering
Acoustic propagation
Acoustic inversion
Characterization of reverberation
Acoustic metamaterials
Acoustics of microfluidic bubbly emulsions
Active sonar performance modeling
Compressive sensing
Acoustic classification
Nonlinear propagation
Underwater acoustic network warfare

Acoustic Simulation, Measurements, and Tactics
Ocean acoustic propagation and scattering
  models
Fleet application acoustic models
High-frequency seafloor and ocean acoustic
  measurements
Riverine acoustics
Distributed sensing networks
Incorporating uncertainty in predictive models
Tactical acoustic simulations and databases
Warfare effectiveness studies and optimization
Environmental assessment and planning tools

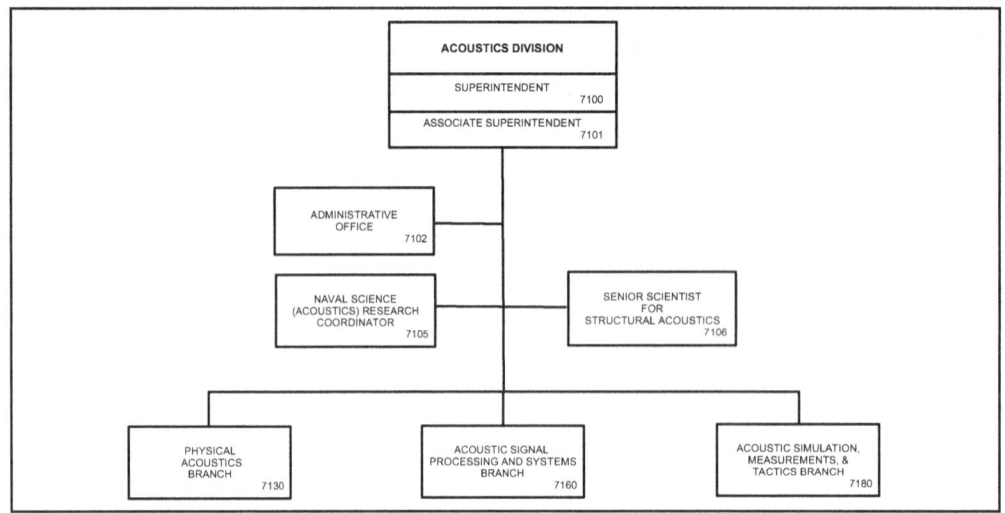

Basic Responsibilities

The Acoustics Division conducts basic and applied research addressing the physics of acoustic signal generation, propagation, scatter, and detection with the objective of improving the strategic and tactical capabilities of the Navy and Marine Corps in the ocean and land operational environment. The Division's scientists and engineers perform collaborative research with scientists affiliated with national and international academic, private, and governmental research organizations. The Division's research spans classical and quantum physics, signal processing, the impact of fluid dynamics on the oceans sound speed field, the propagation and scatter of acoustic signals in the ocean and land environments, structural and physical acoustics including the development of MEMS and nanotechnology based sensors, and the application of networked unmanned underwater vehicles and associated sensors to the Navy's ASW, MCM, and ISR missions.

Personnel: 77 full-time civilian

Key Personnel

Title	Code
Superintendent, Acoustics Division	7100
Associate Superintendent	7101
Administrative Officer	7102
Naval Science (Acoustics) Research Coordinator	7105
Senior Scientist for Structural Acoustics	7106
Head, Physical Acoustics Branch	7130
Head, Acoustic Signal Processing and Systems Branch	7160
Head, Acoustic Simulation, Measurements, and Tactics Branch	7180

Point of contact: Code 7100, (202) 767-3482

Remote Sensing Division

Code 7200
Research Activity Areas

Remote Sensing
Sensors
  SAR
  Imaging radar
  Passive microwave imagers
  CCDs and focal plane arrays
  Thermal IR cameras
  Fabry-Perot spectrometers
  Imaging spectrometers
  Radio interferometers
  Optical interferometers
  Adaptive optics
  Lidar
  Spaceborne and airborne systems
Research Areas
  Radiative transfer modeling
  Coastal oceans
  Marine ocean boundary layer
  Polar ice
  Middle atmosphere
  Global ocean phenomenology
  Environmental change
  Ocean surface wind vector
  Soil moisture
  Ionosphere
  Data assimilation

Astrophysics
Optical interferometry
Radio interferometry
Fundamental astrometry and reference frames
Fundamental astrophysics
Star formation
Stellar atmospheres and envelopes
Interstellar medium,
  interstellar scattering
  pulsars
Low-frequency
  astronomy

Physics of Atmospheric/Ocean Interaction
Mesoscale, fine-structure, and microstructure
Aerosol and cloud physics
Mixed layer and thermocline applications
Sea-truth towed instrumentation techniques
Turbulent jets and wakes
Nonlinear and breaking ocean waves
Stratified and rotating flows
Turbulence modeling
Boundary layer hydrodynamics
Marine hydrodynamics
Computational hydrodynamics

Imaging Research/Systems
Remotely sensed signatures analysis/simulation
Real-time signal and image processing
  algorithm/systems
Image data compression methodology
Image fusion
Automatic target recognition
Scene/sensor noise characterization
Image enhancement/noise reduction
Scene classification techniques
Radar and laser imaging systems studies
Coherent/incoherent imaging sensor exploitation
Numerical modeling simulation
Environmental imagery analysis

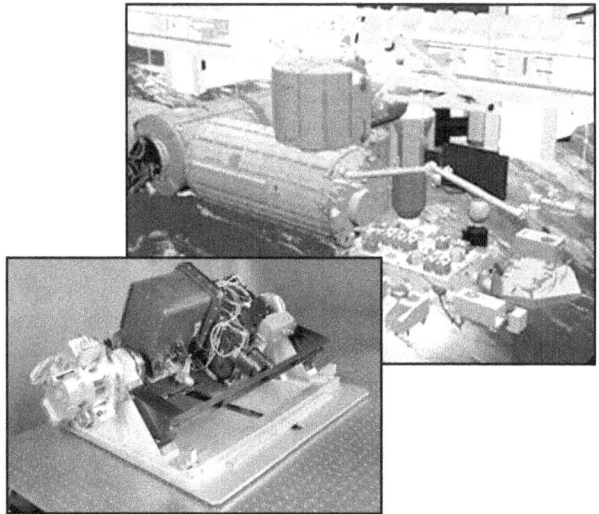

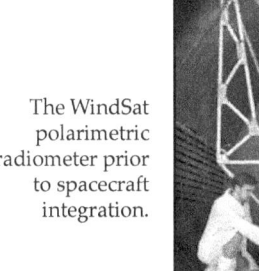
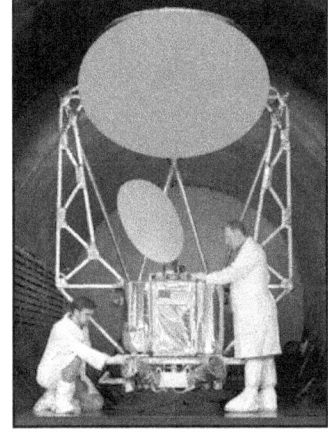

The WindSat polarimetric radiometer prior to spacecraft integration.

The Hyperspectral Imager for the Coastal Ocean, or HICO, is optimized to image the coastal ocean and adjacent land in 128 contiguous color bands. This spectral data is used to develop maps of water depth, water optical properties, land vegetation, and soil bearing strength. HICO was deployed to the International Space Station in September 2009, providing scientific imagery of varied coastal types worldwide.

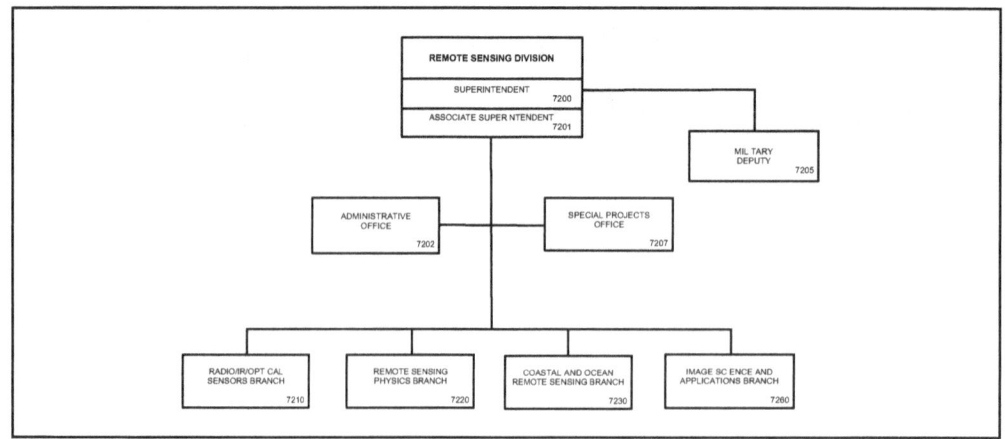

Basic Responsibilities

The Remote Sensing Division is the Navy's center of excellence for remote sensing research and development, conducting a program of basic research, science, and applications aimed at the development of new concepts for sensors and imaging systems for objects and targets on the Earth, in the near-Earth environment, and in deep space. The research, both theoretical and experimental, deals with discovering and understanding the basic physical principles and mechanisms that give rise to target and background emission and to absorption and emission by the intervening medium. The accomplishment of this research requires the development of sensor systems technology. This development effort includes active and passive sensor systems to be used for the study and analysis of the physical characteristics of phenomena that give rise to naturally occurring background radiation, such as that caused by the Earth's atmosphere and oceans, as well as man-made or induced phenomena, such as ship/submarine hydrodynamic effects. The research also includes theory, laboratory, and field experiments leading to ground-based, airborne, and space-based systems for use in such areas as environmental remote sensing (including improved meteorological support systems for the operational Navy), astrometry, astrophysics, surveillance, and nonacoustic ASW. Special emphasis is given to developing space-based platforms and exploiting existing space systems.

Personnel: 97 full-time civilian

Key Personnel

Title	Code
Superintendent, Remote Sensing Division	7200
Associate Superintendent	7201
Administrative Officer	7202
Military Deputy	7205
Head, Radio/Infrared/Optical Sensors Branch	7210
Head, Remote Sensing Physics Branch	7220
Head, Coastal and Ocean Remote Sensing Branch	7230
Head, Image Science and Applications Branch	7260

Point of contact: Code 7200, (202) 767-3391

Code 7300
Research Activity Areas

Ocean Dynamics and Prediction

Circulation
  Global resolution of circulation and meso-scale fields
  Littoral circulation at the coast, bays, and estuaries
  Satellite observation processing and assimilation
  UUV adaptive sampling
  Observation system simulation experiments
  Ice volume and ice drift
  Tidal currents and heights
Surface effects
  Surface wave effects globally and into bays
  Wave breaking
  Mixed layer dynamics
  Swell propagation and dynamics
  Phase averaged wave evolution
  Phase resolved wave dynamics
Nearshore
  Wave breaking at the shore
  Rip currents at the shore
  Tidal currents and heights into rivers
  Nonlinear wave interaction
  Sensor deployment optimization
Acoustic effects
  Sound speed variation for acoustic propagation
  Internal waves, solitons, and bores for beam focusing
  Wave bubble entrainment and noise generation

Ocean Sciences

Dynamical processes
  Optical turbulence
  Biological sensing and modeling
  Optical thin layers
  Coastal current systems
  Waves and bubbles
Coupled systems
  Air/ocean/acoustic coupling
  Coupled bio/optical/physical processes
  Coupled physical/sediment processes
Remote sensing applications
  3D optical profiling
  Color/hyperspectral signatures
  Ocean optics
  Sea surface salinity
Microbiologically influenced corrosion
  Metal-microbe interaction

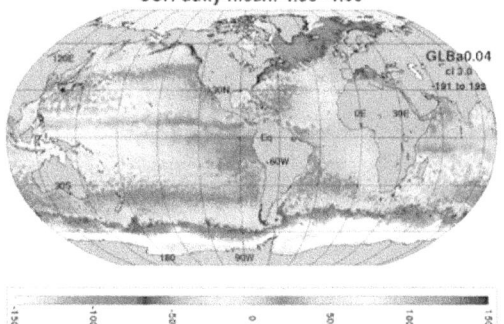

Global sea surface height from the 1/25° Hybrid Coordinate Ocean Model (HYCOM) including ice cover.

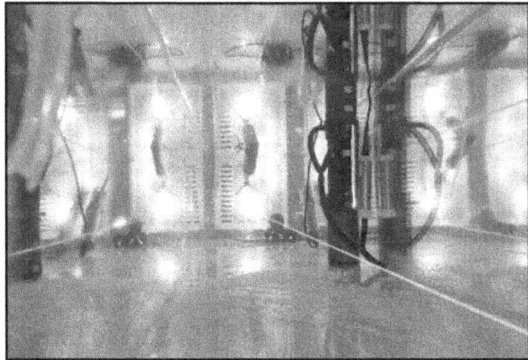

Rayleigh Bernard Convective Tank provides a controlled environment capable of generating turbulent microstructures at various repeatable intensities.

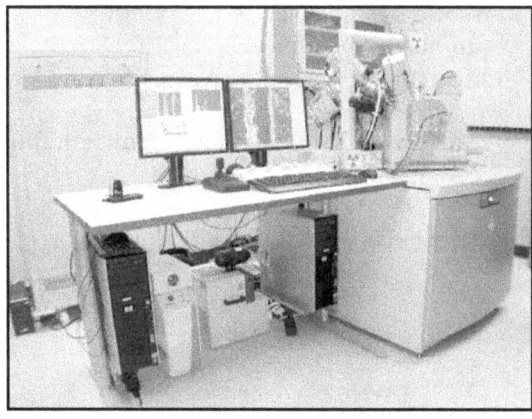

Environmental scanning electron microscope with focused ion beam (ESEM/FIB) coupled with an energy dispersive X-ray detector.

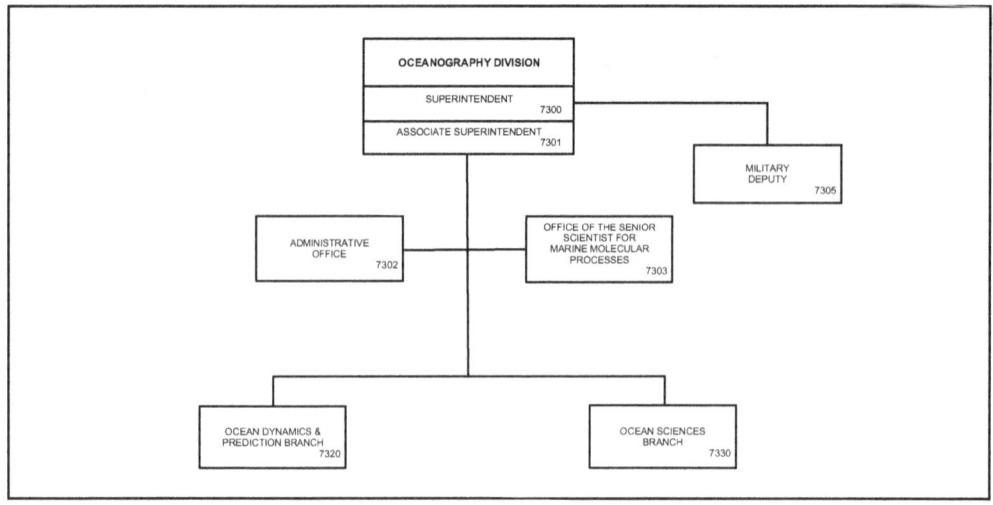

Basic Responsibilities

The Oceanography Division conducts basic and applied research in description and modeling of biological, physical, and dynamical processes in open ocean, regional, and littoral areas; in exploitation of satellite, airborne, and in situ sensors for environmental characterization; and in investigation and application of microbial processes to Navy problems. The oceanographic research is both theoretical and experimental in nature and is focused on understanding and modeling ocean, coastal, and littoral area hydro/thermodynamics, circulation, waves, ice dynamics, air-sea exchange, optics, and small and microscale processes. Analytical methods and algorithms are developed to provide quantitative retrieval of geophysical parameters of Navy interest from state-of-the-art sensor systems. The Division work includes analysis of biological processes that mediate and control optical properties of the oceans, coastal, and littoral regions, and microbially induced corrosion/metal-microbe interaction. The Division programs are designed to be responsive to and to anticipate Naval needs. Transition of Division products to the DoD, Navy systems developers, operational Navy, and civilian (dual use) programs is a primary goal. The Division's programs are coordinated and interactive with other NRL programs and activities, ONR's research programs, and other government agencies involved in oceanographic activities. The Division also collaborates and cooperates with scientists from the academic community and other U.S. and foreign laboratories.

Personnel: 86 full-time civilian; 1 military

Key Personnel

Title	Code
Superintendent, Oceanography Division	7300
Associate Superintendent	7301
Administrative Officer	7302
Office of the Senior Scientist for Marine Molecular Processes	7303
Military Deputy	7305
Head, Ocean Dynamics and Prediction Branch	7320
Head, Ocean Sciences Branch	7330

Point of contact: Code 7301, (228) 688-4704; DSN 828-4704

Code 7400
Research Activity Areas

Marine Geology

Sedimentary processes
Sediment microstructure
Pore fluid flow
Diapirism, volcanism, faulting, mass movement
Biogenic and thermogenic methane
Hydrate distribution, formation, and dissociation
Small-scale granular/fluid dynamics

Marine Geophysics

Seismic wave propagation
Physics of low-frequency acoustic propagation
Acoustic energy interaction with topography and
 inhomogeneities
Gravimetry and geodesy
Geomagnetic modeling

Marine Geotechnique

Acoustic seafloor characterization
Geoacoustic modeling
Geotechnical properties and behavior of sedi-
 ments
Measurement and modeling of high-frequency
 acoustic propagation and scattering
Mine burial processes
Marine biogeochemistry
 Animal-microbe-sediment interactions
 Early sediment diagenesis
Biomineralization of palladium species
Physics-based and numerical modeling of
 sediment strength

Geospatial Sciences and Technology

Digital database design
Digital product analysis and standardization
Data compression techniques and exploitation
Hydrographic survey techniques
Bathymetry extraction techniques from remote and
 acoustic imagery
Modeling of nearshore morphodynamics
Geospatial portal design with 2D and 3D interfaces
Characterization of the littoral from airborne
 platforms

In Situ and Laboratory Sensors

High-resolution subseafloor 2D and 3D seismic
 imaging
Laser/hyperspectral bathymetry/topography
Swath acoustic backscatter imaging
Sediment pore water pressure, permeability, and
 undrained shear strength
Compressional and shear wave velocity and
 attenuation
Airborne geophysics, gravity, and magnetics
Seafloor magnetic fluctuation
Sediment microfabric change with pore fluid
 and/or gas change
Instrumented mine shapes
Bottom currents and pressure fluctuations

In the Marine Geosciences Division, scientists perform laboratory experiments with a small oscillatory flow tunnel (S-OFT) to study the formation and migration of sand ripples. Rippled sand beds are ubiquitous on the seafloor in shallow water. Understanding the complex response of the seafloor to forcing from surface waves and currents is important for Naval operations from amphibious landings to mine warfare. Shown in the image is the S-OFT including a mounted laser and four high-speed video cameras to perform tomographic particle image velocimetry (Tomo-PIV) measurements, which estimate the three-dimensional fluid velocity in a volume up to 10 cm^3. The upper inset is a picture of a sand ripple formed using a bimodal distribution of sand where the smaller sand particles are darker and the larger sand particles are lighter in color. The lower inset is a profile image of a sand ripple from the same experiment where the sorting processes between large and small grains have formed visible strata. Ripple migration is from right to left in both inset images.

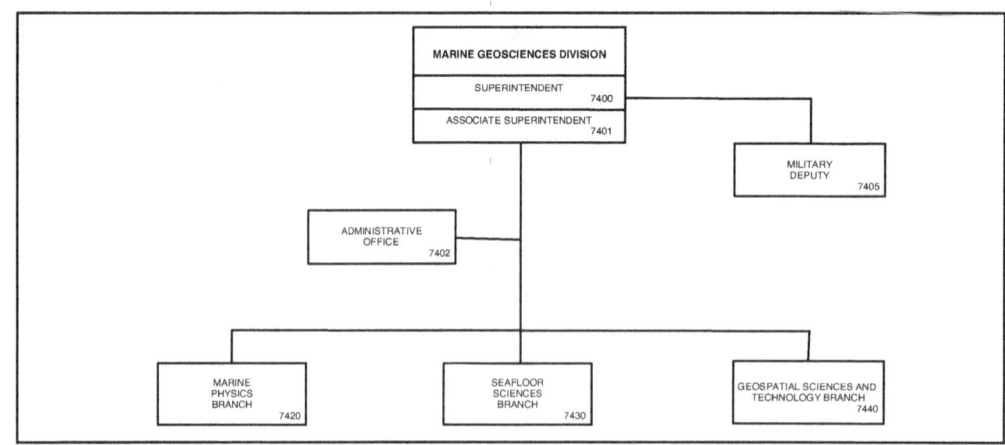

Basic Responsibilities

The Marine Geosciences Division conducts a broadly based, multidisciplinary program of scientific research, advanced technology development, and applied research in marine geosciences, geodesy, geospatial information, and related technologies. This includes investigations of basic processes within ocean basins, littoral regions and adjacent land areas, and arctic regions; development of models, sensors, and techniques; and the exploitation of this knowledge and technology to enhance Navy and Marine Corps systems, plans, and operations, and to meet national needs.

As the Navy's subject matter expert in the areas of Geospatial Information and Services (GI&S), the Division provides vital technical support to the Oceanographer/Navigator of the Navy, CNO, N2/N6F5, the National Geospatial-Intelligence Agency (NGA) and the Tri-Service Community. NRL also contributes to the development of leading-edge geospatial technology by reviewing emerging GI&S standards and products.

Close coordination and interactions with the Commander, Naval Meteorology and Oceanography Command, Naval Oceanographic Office, CNO, Office of Naval Research (ONR), Systems Commands, Warfare Centers, NGA, and the other DoD and national organizations are essential to the success of Division programs, with transition of Division technology to systems developers and to the operational Navy a primary goal. The Division program is coordinated and interactive with other NRL programs and activities, ONR's Research Program Department, NOAA, USGS, NSF, and other government agencies involved in seafloor activities. The Division collaborates and cooperates with scientists from the academic community, other U.S. and foreign laboratories, and industry.

Personnel: 63 full-time civilian; 2 military

Key Personnel

Title	Code
Superintendent, Marine Geosciences Division	7400
Associate Superintendent	7401
Administrative Officer	7402
Military Deputy	7405
Head, Marine Physics Branch	7420
Head, Seafloor Sciences Branch	7430
Head, Geospatial Sciences and Technology Branch	7440

Point of contact: Code 7402, (228) 688-4660; DSN 828-4660

Marine Meteorology Division

Code 7500
Research Activity Areas

Atmospheric Dynamics and Prediction *
- Global to tactical scale
- Deterministic and probabilistic
- Large eddy simulation
- Boundary layer
- Land surface
- Coastal
- Arctic
- Urban effects
- Massively parallel computing
- Coupled ocean/atmosphere
- Tropical cyclones
- Aerosols
- Topographically forced flow
- Predictability
- Ensembles design
- Advanced numerical methods

Data Assimilation
- Hybrid techniques
- 3D and 4D variational analysis
- Ensemble Transform Kalman Filter (ETKF)
- Quality control and bias correction
- Tropical cyclone initialization
- Remotely sensed data assimilation
- Adjoint applications
- Direct radiance assimilation
- Radar data assimilation
- Targeted observations
- Data selection techniques
- Aerosol assimilation
- UAV data assimilation

Tactical Environmental Support
- Rapid environmental assessment
- Through-the-sensor measurements
- Atmospheric impact on weapons systems
- Chem-bio transport and dispersion
- Data fusion
- Nowcasting
- Visualization
- Expert systems
- Aviation risk assessment

Atmospheric Physics
- Air-sea interaction
- Cloud and aerosol microphysics
- Radiative transfer
- Aerosol characterization
- Tropical cyclone structure

Satellite Data/Imagery
- Automated classification of cloud properties
- Sensor calibration/validation
- Satellite imagery analysis and enhancement
- Multisensor data fusion
- Tropical cyclone characterization
- Dust/aerosols
- Rain rate and snow cover
- Nighttime environmental analysis
- JPSS preparation
- Tactical meteorology

Decision Aids
- Refractivity/ducting
- Ceiling/visibility
- Fog/turbulence/icing
- Atmospheric acoustics
- EM/EO propagation effects
- Tropical cyclones/consensus forecasts
- Nuclear/chemical/biological transport and dispersion
- Port studies
- Typhoon havens
- Forecaster handbooks
- Quantification of uncertainty
- Counter-piracy guidance
- Tropical cyclone sortie guidance
- Forecast difficulty guidance
- Ship wind and wave limits
- Optimal ship routing – fuel savings

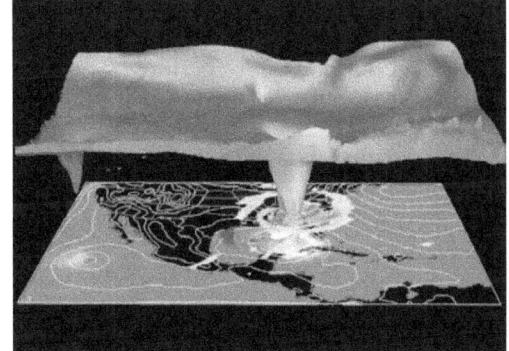

A 3D depiction of forecast sensitivity based on a COAMPS model forecast of Hurricane Katrina, obtained using the model's adjoint and tangent linear model system. The sea-level pressure (white contours) and 10 m wind speed are shown at the surface. The sensitivity of the energy in a box surrounding Katrina to the previous 24-h model vorticity at 2.5 km is shown elevated above the surface. The 3D surface corresponding to the equivalent potential temperature of 340 K, shaded by wind speed, is also displayed.

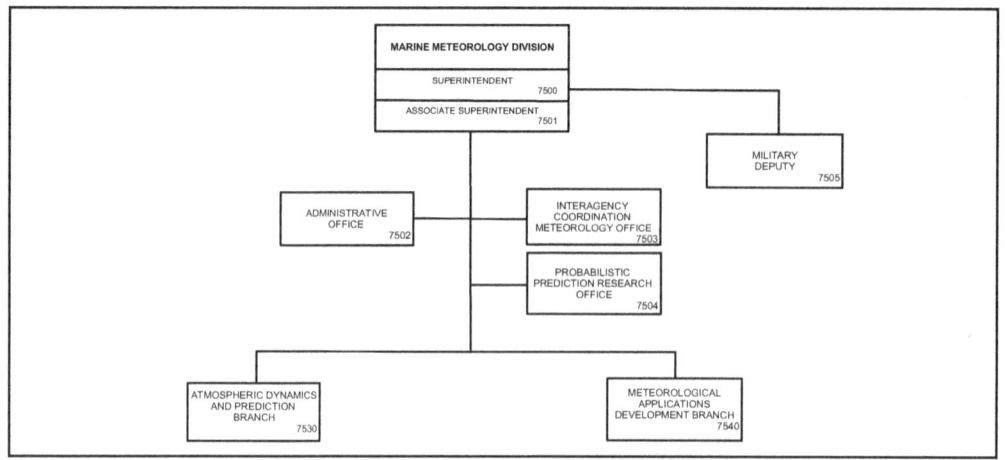

Basic Responsibilities

The Marine Meteorology Division conducts a basic and applied research and development program designed to improve scientific understanding of atmospheric processes that impact Fleet operations and to develop automated systems that analyze, simulate, predict, and interpret the structure and behavior of these processes and their effect on naval weapons systems. Basic and applied research includes work in air-sea interaction, aerosol and cloud physics, atmospheric turbulence, orographically forced flow, atmospheric predictability, scale interactions observation impact, advanced data assimilation, ensemble prediction, tropical dynamics, and numerical methods. Research and development ranges from development of atmospheric analysis/forecast systems and satellite data products to the development of tactical decision aids for operations support. Interdisciplinary research supports the development of coupled analysis/forecast systems, including components for ocean, wave, land surface, aerosol, chemistry, and middle atmosphere prediction. NRL-Monterey (NRL-MRY) is co-located with the Fleet Numerical Meteorology and Oceanography Center (FNMOC) and has developed and transitioned to FNMOC the data assimilation, global and meso-scale weather forecast models, aerosol prediction systems, and satellite applications products that form the backbone of the Navy's worldwide environmental forecasting capability. Specialties of the Division include numerical weather prediction, data assimilation, tropical cyclones, marine boundary layer processes, aerosols, rapid environmental assessment, environmental decision aids, and satellite data analysis, interpretation, and application.

Personnel: 77 full-time civilian; 1 military

Key Personnel

Title	Code
Superintendent, Marine Meteorology Division	7500
Associate Superintendent	7501
Administrative Officer	7502
Head, Interagency Coordination Meteorology Office	7503
Lead Scientist, Probabilistic Prediction Research Office	7504
Military Deputy	7505
Head, Atmospheric Dynamics and Prediction Branch	7530
Head, Meteorological Applications Development Branch	7540

Point of contact: Code 7500, (831) 656-4721; DSN 878-4721

Code 7600
Research Activity Areas

Geospace Science and Technology

Research to observe, understand, model, and forecast the Earth's operational environment that extends from the lower atmosphere to the magnetopause, in which region both terrestrial and solar effects influence the space environment.

First monolithic Doppler Asymmetric Spatial Heterodyne Spectroscopy (DASH) interferometer. DASH is an innovative, advanced optical technique that can be used to measure winds in the middle and upper atmosphere of Earth and on other planets.

Solar and Heliospheric Physics

Research to develop a fundamental physical understanding of highly variable transient and long-term solar activity; the radiative, plasma, and particulate emissions associated with the activity; and the responses of the heliosphere and the terrestrial magnetosphere to the activity. Relevant empirical data is collected by conceiving, developing, and operating state-of-the-art imaging, spectrometric, and in situ space flight sensors on national and international space missions. Physics-based models are hypothesized, tested with the collected empirical data and computer simulation, and developed.

High Energy Space Environment

Research of energetic particle, γ-ray, and X-ray/ultraviolet environments in space and for other applications of interest to the DoD, homeland security, and national programs, such as detection and surveillance of nuclear materials in terrestrial and space applications.

SECCHI: The Sun-Earth Connection and Heliospheric Investigation instrument suite, shown during testing at NRL, is returning spectacular stereo imagery of the region between the Sun and the Earth.

GLAST launched at 12:05 p.m. EDT on 11 June 2008 from Cape Canaveral Air Force Station on a Delta II 7920-10 rocket. After on-orbit checkout and commissioning, the observatory was renamed the Fermi Gamma-ray Space Telescope in honor of Enrico Fermi.

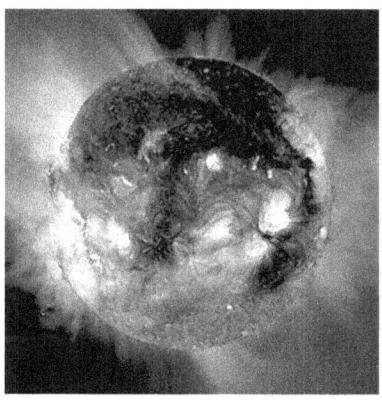

Solar image taken with the Extreme Ultraviolet Imaging Telescope (EIT) on the Solar and Heliospheric Observatory (SOHO) spacecraft. The bright areas are active regions above sunspots, and the dark areas are coronal holes where the open magnetic structure allows the fast solar wind to flow freely out into space.

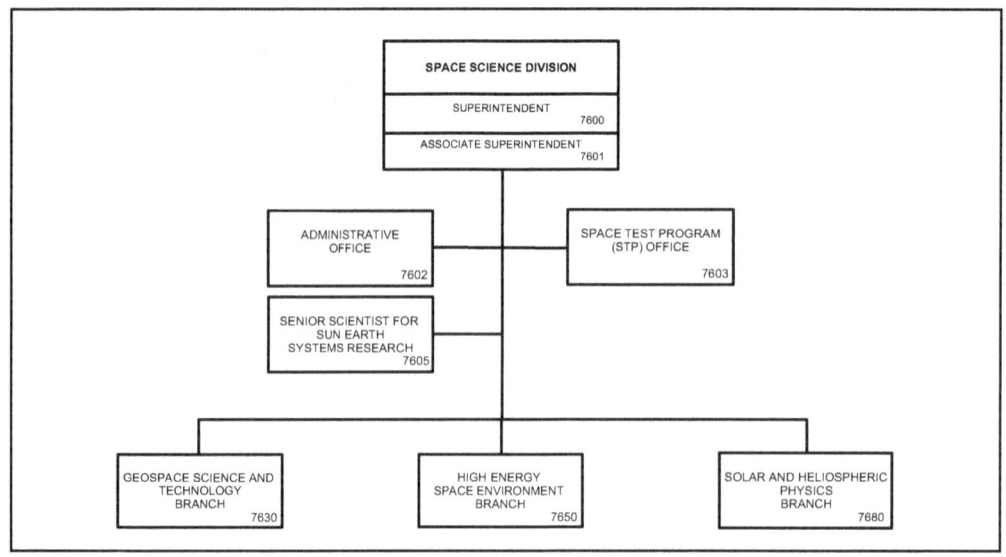

Basic Responsibilities

The Space Science Division conducts a broad-spectrum RDT&E program in solar-terrestrial physics, astrophysics, upper/middle atmospheric science, and astronomy. Instruments to be flown on satellites, sounding rockets and balloons, and ground-based facilities and mathematical models are conceived and developed. Researchers apply these and other capabilities to the study of the atmospheres of the Sun and Earth, including solar activity and its effects on the Earth's ionosphere, upper atmosphere, and middle atmosphere; laboratory astrophysics; and the unique physics and properties of celestial sources. The science is important to orbital tracking, radio communications, and navigation that affect the operation of ships and aircraft, utilitization of the near-space and space environment of the Earth, and the fundamental understanding of natural radiation and geophysical phenomena.

Personnel: 81 full-time civilian; 1 military

Key Personnel

Title	Code
Superintendent, Space Science Division	7600
Associate Superintendent	7601
Administrative Officer	7602
Space Test Program Officer, Kirtland AFB, NM	7603
Senior Scientist for Sun-Earth Systems Research	7605
Head, Geospace Science and Technology Branch	7630
Head, High-Energy Space Environment Branch	7650
Head, Solar and Helioshperic Physics Branch	7680

Point of contact: Code 7602, (202) 767-3248

**Naval Center
for Space
Technology**

NAVAL CENTER FOR SPACE TECHNOLOGY

Code 8000

In its role to preserve and enhance a strong space technology base and provide expert assistance in the development and acquisition of space systems that support naval missions, the Naval Center for Space Technology performs basic and applied research through advanced development in all areas of interest to the Navy space program. The Center develops spacecraft, systems using these spacecraft, and ground command and control stations. Principal functions of the Center include understanding and clarifying requirements, recognizing and prosecuting promising research and development, analyzing and testing systems to quantify their capabilities, developing operational concepts that exploit new technical capabilities, performing system engineering to allocate design requirements to subsystems, and performing engineering development and initial operation to test and evaluate selected spacecraft subsystems and systems. The Center is a focal point and integrator for those divisions at NRL whose technologies are used in space systems. The Center also provides systems engineering and technical direction assistance to system acquisition managers of major space systems. In this role, technology transfer is a major goal and motivates a continuous search for new technologies and capabilities and the development of prototypes that demonstrate the integration of such technologies.

M r. P.G. Wilhelm was born in New York City. He attended Purdue University, where he received a B.S.E.E. degree in 1957. By 1961, he had completed all the course work for an M.S.E. degree from George Washington University.

From 1957 to 1959, Mr. Wilhelm served as an electrical engineer with Stewart Warner Electronics where he was assigned to a project to redesign the UPM-70, a Navy radar test set. In March 1959, he joined the Naval Research Laboratory as an electrical scientist in the Electronics Division. In December 1959, he joined the Satellite Techniques Branch. In 1961, he became Head of the Satellite Instrument Section; in 1965, he became Head of the Satellite Techniques Branch; and in 1974, Head of the Spacecraft Technology Center. In these positions, he performed satellite system design, equipment development, environmental testing, launch operations, and orbital data handling. In 1981, he was named Superintendent of the Space Systems and Technology Division, the Navy's principal organization, or lead laboratory, for space. He is credited with contributions in the design, development, and operation of more than 100 scientific and Fleet-support satellites. He has been awarded five patents. In October 1986, he was appointed Director of the newly established Naval Center for Space Technology. The Center's mission is to "preserve and enhance a strong space technology base and provide expert assistance in the development and acquisition of space systems which support naval missions."

Mr. Wilhelm has been recognized with numerous awards including the Navy's Meritorious Civilian Service Award, the DoD Distinguished Civilian Service Award, the Presidential Meritorious Executive Award, the Presidential Distinguished Rank Award, the Institute of Electrical and Electronics Engineers Aerospace and Electronic Systems Group Man of the Year Award, the NRL E.O. Hulburt Annual Science and Engineering Award, the Dexter Conrad Award, the Rotary National Stellar Award, the NRL Lifetime Achievement Award, and in May 1999, Mr. Wilhelm received the American Institute of Aeronautics and Astronautics (AIAA) Goddard Astronautics Award. He also has been elected a Fellow of the Washington Academy of Sciences and a Fellow of the American Institute of Aeronautics and Astronautics, and was elected to the National Academy of Engineering. Mr. Wilhelm is also the first recipient of the R.L. Easton Award for excellence in engineering.

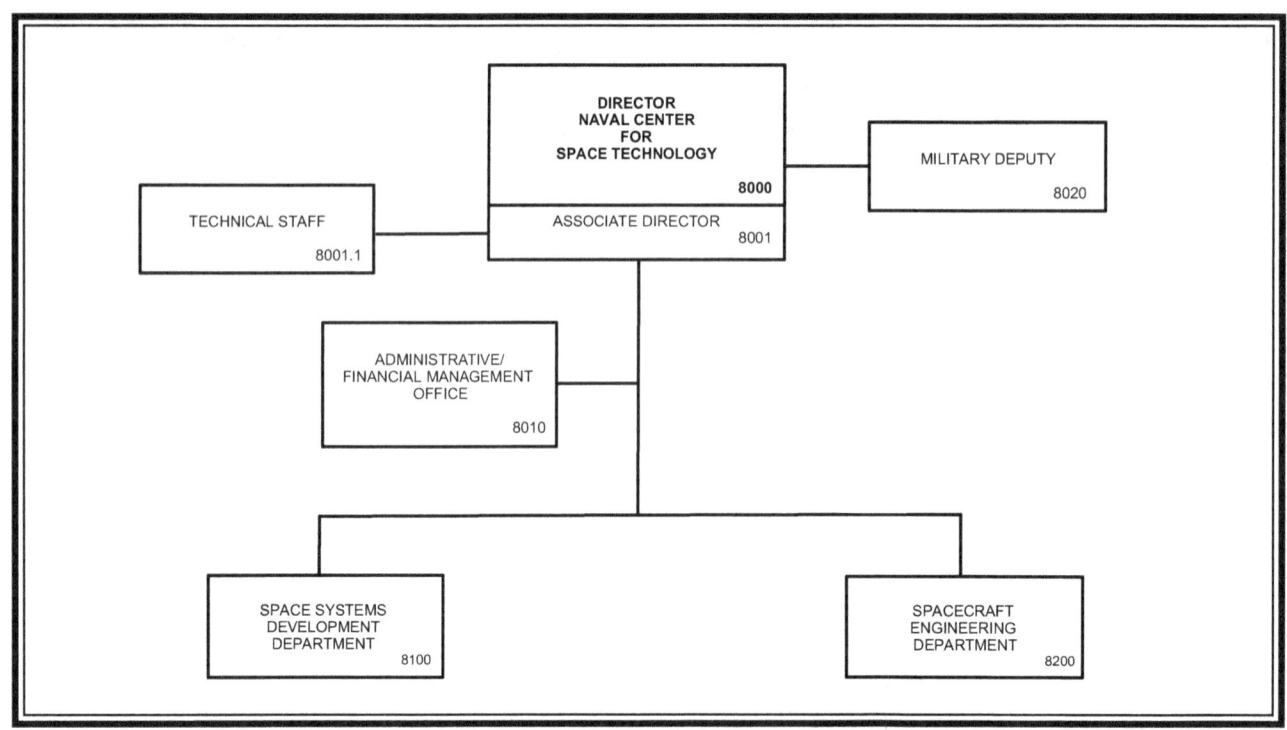

Key Personnel

Title	Code
Director, Naval Center for Space Technology	8000
Associate Director	8001
Technical Staff	8001.1
Head, Administrative/Financial Management Office	8010
Military Deputy	8020
Superintendent, Space Systems Development Department	8100
Superintendent, Spacecraft Engineering Department	8200

Point of contact: Code 8010, (202) 767-6550

Space Systems Development Department

Code 8100
Research Activity Areas

Advanced Space/Airborne/Ground Systems Technologies
Space systems architectures and requirements
Advanced payloads and optical communications
Controllers, processors, signal processing, and VLSI data management systems and equipment
Embedded algorithms and software
Satellite laser ranging

Astrodynamics
Precision orbit estimation
Onboard autonomous navigation
Onboard orbit propagation
GPS space navigation
Satellite coverage and mission analysis
Geolocation systems
Orbit dynamics
Interplanetary navigation

Command, Control, Communications, Computers, Intelligence, Surveillance, and Reconnaissance
Communications theory and systems
Satellite ground station engineering and implementation
Transportable and fixed ground antenna systems
High-speed fixed and mobile ground data collection, processing, and dissemination systems
Tactical communication systems

Space and Airborne Payload Development
Space and airborne system payload concept definition, design, and implementation including hardware and software
Detailed electrical/electronic design of electronic and electromechanical payload and systems and components
Design and verification of real-time embedded multiprocessor software
Payload antenna systems
Space and airborne payload fabrication, test, and integration
Launch and on-orbit payload support

Laser Communications Research
Ship-to-ship laser communications
Space-to-ground laser communications
Satellite laser ranging for precise orbit determination

Space and Airborne Mission Development
Mission development and requirements definition
Systems engineering and analysis
Concepts of operations and mission simulations
Mission evaluation and performance assessments

Precision Navigation and Time
Advanced navigation satellite technology
Precise Time and Time Interval (PTTI) technology
Atomic time/frequency standards/instrumentation
Passive and active ranging techniques
Precision tracking of orbiting objects from space/ground
National and International standards for timekeeping/ Universal Coordinated Time/UTC (NRL)

SEALINK Advanced Analysis (S2A) provides global, persistent, cooperative and non-cooperative maritime vessel tracking awareness and information that is valuable to intelligence analysts, joint warfighters, senior decision makers, and interagency offices within the SCI community.

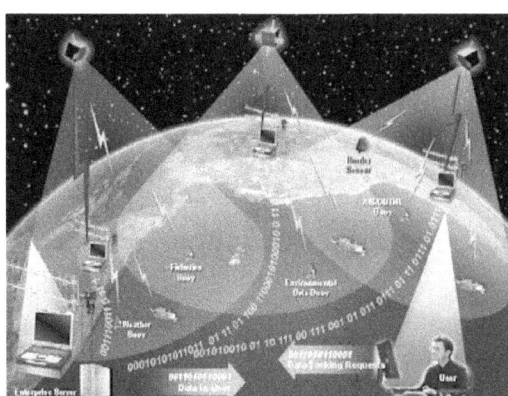

The Global Awareness and Data Extraction International Satellite (GLADIS) is a system of 30 satellites designed to achieve expanded global situational awareness and information sharing.

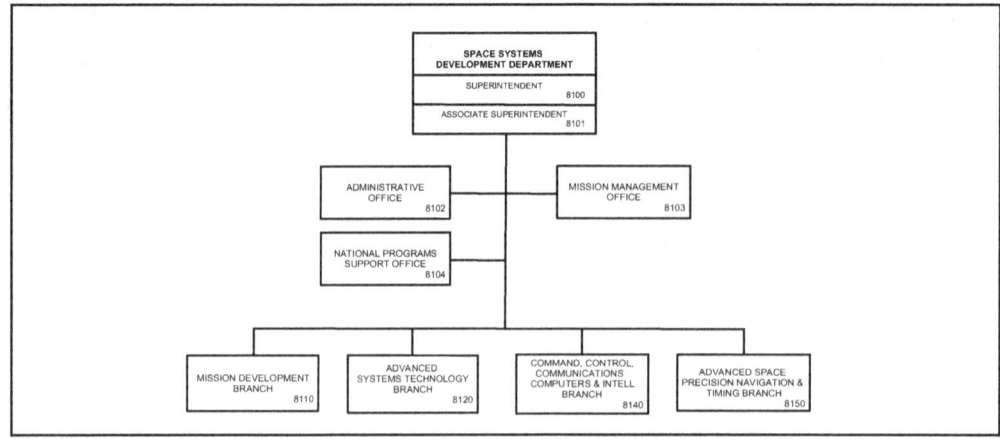

Basic Responsibilities

The Space Systems Development Department (SSDD) is the space and ground support systems research and development organization of the Naval Center for Space Technology. The primary objective of the SSDD is to develop command, control, communications, computers, and intelligence, surveillance, and reconnaissance (C4ISR) hardware and software solutions to space, airborne, and ground applications to respond to Navy, DoD, and national mission requirements with improved performance, capacity, reliability, efficiency, and/or life cycle cost. The Department must derive system requirements from the mission, develop architectures in response to these requirements, and design and develop systems, subsystems, equipment, and implementation technologies to achieve the optimized, integrated operational space, airborne, and ground system. These development responsibilities extend across the entire space/airborne/ground spectrum of hardware, software, and advanced technologies, including digital processing and control, analog systems, power, communications, payload command and telemetry, radio frequency, optical, payload, and electromechanical systems, as well as systems engineering.

Personnel: 126 full-time civilian; 1 part-time civilian; 23 student civilian; 1 intermittent civilian

Key Personnel

Title	Code
Superintendent, Space Systems Development Department	8100
Associate Superintendent	8101
Administrative Officer	8102
Head, Mission Management Office	8103
Head, National Programs Support Office	8104
Head, Mission Development Branch	8110
Head, Advanced Systems Technology Branch	8120
Head, Command, Control, Communications, Computers, and Intelligence Branch	8140
Head, Advanced Space Precision Navigation and Timing Branch	8150

Point of contact: Code 8102, (202) 767-0432

Spacecraft Engineering Department

Code 8200
Research Activity Areas

Design, Test, and Processing
Design, fabrication, and testing of spacecraft and hardware

Preliminary and detailed design, fabrication, testing, and integration onto launch vehicle

Systems engineering for new spacecraft proposals

Start-to-finish responsibility for NCST spacecraft mechanical systems

Space Mechanical Systems Development
Research and development in spacecraft technology

Conceptual design trade studies

Integrated engineering design and analysis

Structural and thermal design and analysis

Development and transition of prototype hardware

Development and integration of experimental payloads

Mission integration and development

Control Systems
Attitude determination and control systems

Precision pointing

Optical line-of-sight stabilization

Propulsion systems

Precision cleaning and component testing

Propellent and pressurization systems

Hydraulic and pneumatics control

Test systems and services

Analytical design and mission planning

Navigation, tracking, and orbit dynamics

Expert systems

Flight operations support

Computer simulation

Computer animation

Robotics systems engineering

Proximity operations

Autonomous servicing

Autonomous inspection

End effector design

Compliance control

Trajectory planning

Machine vision

Fault detection, isolation, and recovery

Space Electronic Systems Development
Space system concept definition, design, and implementation including hardware and software

Detailed electrical/electronic design of electronic and electromechanical systems and components

Implementation of real-time flight software and embedded command, control, and telemetry software

Design and verification of real-time embedded multi-processor software

Spacecraft antenna systems

Space systems fabrication, test, and integration

Launch and on-orbit support

Space test systems and electronic launch support equipment

Space TT&C and control systems

Space communication systems

The Space Robotics Laboratory employs two six-degree-of-freedom robotic manipulators to perform realistic orbital and attitude motion simulations for proximity operations of spacecraft. This facility enables hardware-in-the-loop testing of machine vision systems, capture mechanisms and autonomous guidance, navigation, and control algorithms. The resulting technologies will benefit future DoD space missions involving autonomous rendezvous and capture.

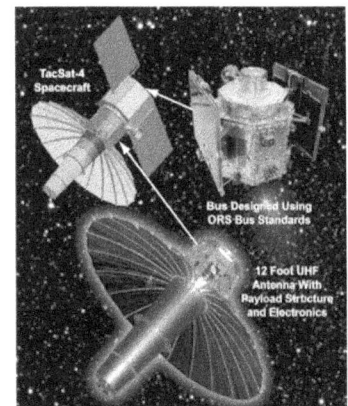

TacSat-4 is a Navy-led joint mission to provide operationally relevant capabilities and enable Operationally Responsive Space (ORS). TacSat-4 provides 10 ultra high frequency channels that can be used for any combination of communications, data exfiltration, or Blue Force tracking. Notably, TacSat-4 provides communications on-the-move with legacy radios and provides a wideband "MOUS-like" channel for early testing. The unique orbit augments geosynchronous communications by allowing near-global, but not continuous, coverage including the high latitudes. TacSat-4 also advances ORS development areas including spacecraft bus standards, long dwell orbits, dynamic tasking, and net-centric operations. TacSat-4 launched in 2011.

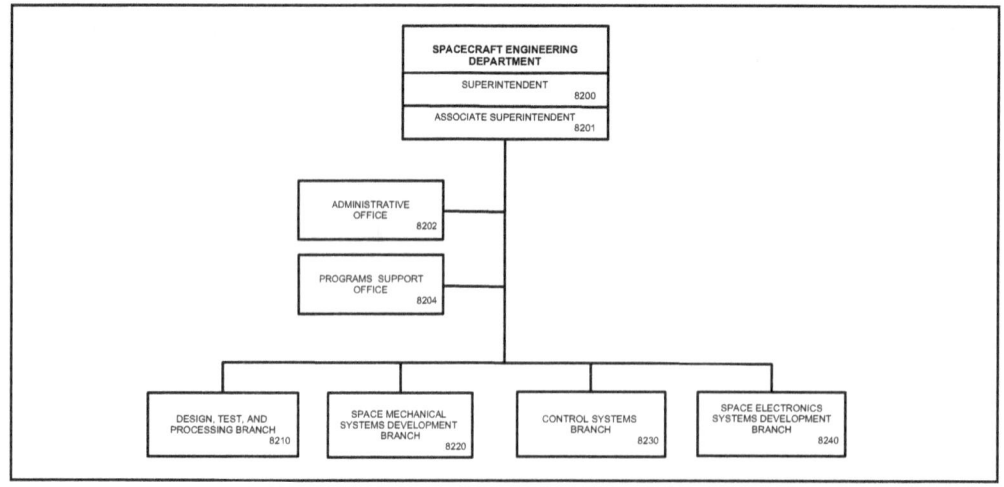

Basic Responsibilities

The Spacecraft Engineering Department (SED) is the focal point for the Navy's capability to design and build spacecraft. Activities range from concept and feasibility planning to on-orbit IOC for NRL's space systems.

The SED provides spacecraft bus expertise for the Navy and maintains an active in-house capability to develop satellites; manages Navy space programs through engineering support and technical direction; in concert with the Space Systems Development Department, designs, assembles, and tests spacecraft and space experiments, including all aspects of space, launch, and ground support; analyzes and designs structures, mechanisms, and a variety of control systems, including attitude, propulsion, reaction, and thermal; integrates satellite designs, launch vehicles, and satellite-to-boost stages; functions as a prototype laboratory to ensure that designs can be transferred to industry and incorporated into subsequent satellite hardware builds; and consults with the Navy Program Office on technical issues involving spacecraft architecture, acquisition, and operation.

Personnel: 128 full-time civilian; 2 part-time civilian; 26 student civilian

Key Personnel

Title	Code
Superintendent, Spacecraft Engineering Department	8200
Associate Superintendent	8201
Administrative Officer	8202
Head, Programs Support Office	8204
Head, Design, Test, and Processing Branch	8210
Head, Space Mechanical Systems Development Branch	8220
Head, Control Systems Branch	8230
Head, Space Electronics Systems Development Branch	8240

Point of contact: Code 8202, (202) 767-6412

Technical Output, Fiscal, and Personnel Information

Publications, Presentations, and Patents

The Navy continues to be a pioneer in science and engineering developments and a leader in applying these advancements to military requirements. The primary means of informing the scientific and engineering community of the advances made at NRL is through the Laboratory's technical output—reports, articles in scientific journals, contributions to books, papers presented to scientific societies and topical conferences, patents, and inventions.

The figures for calendar years 2010 and 2011 presented below represent the output of NRL facilities in Washington, DC; Bay St. Louis, Mississippi; and Monterey, California.

In 1986, Congress enacted the Federal Technology Transfer Act in an effort to encourage the commercial use of technology developed in Federal laboratories. The Act allows Government inventors and the laboratories where they work to share the royalties generated by commercial licensing of their inventions. Also, the Act encourages the establishment of Cooperative Research and Development Agreements (CRADAs) between laboratories such as NRL and non-Federal entities such as state and local governments, universities, and business corporations. Such cooperative R&D agreements can include the allocation in advance of patent rights on any inventions made under the joint research effort.

The 1986 Act has given additional impetus to the Laboratory's efforts to patent important inventions arising out of its various research programs.

Calendar Year 2010

Type of Contribution	Unclassified	Classified	Total
Articles in periodicals, chapters in books, and papers in published proceedings	1502	0	1502
Oral Presentations	2181	0	2181
NRL Formal Reports	6	6	12
NRL Memorandum Reports	67	8	75
Books	2	0	2
Patents granted	51	0	51
Statutory Invention Registrations (SIRs)	0	0	0

Calendar Year 2011

Type of Contribution	Unclassified	Classified	Total
Articles in periodicals, chapters in books, and papers in published proceedings	1398	0	1398*
Oral Presentations	2301	0	2301
NRL Formal Reports	6	5	11
NRL Memorandum Reports	51	4	55
Books	1	0	1
Patents granted	87	1	88
Statutory Invention Registrations (SIRs)	0	0	0

*This is a provisional total based on information available to the Ruth H. Hooker Research Library on August 1, 2012. Total includes refereed and non-refereed publications.

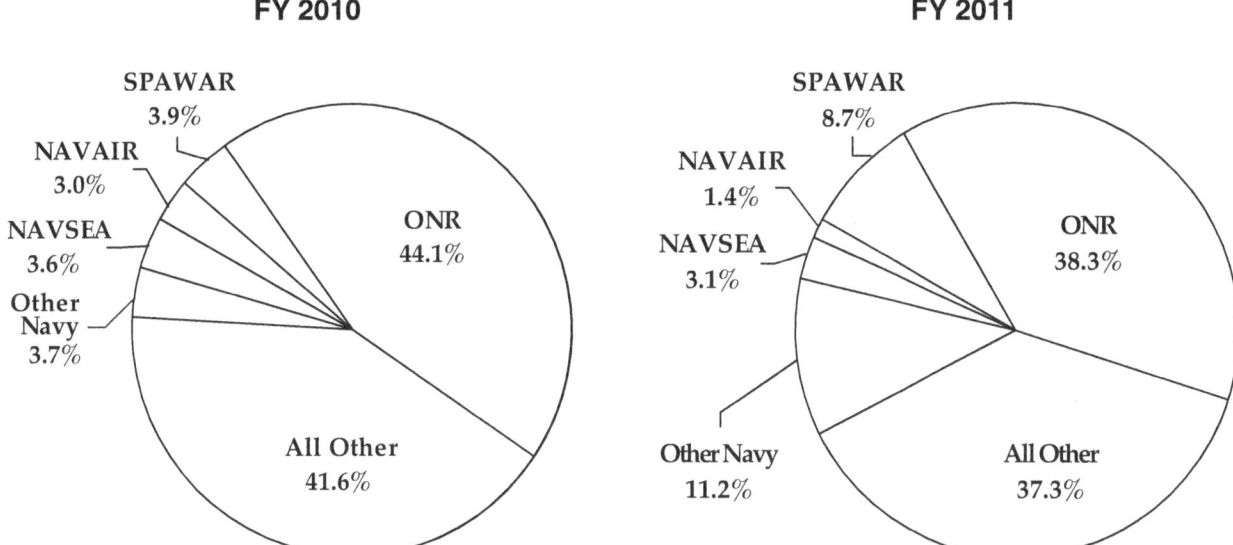

FY 2010

FY 2011

FY 2010
Source of Funds (%)

FY 2010	Reimbursable	$M Direct Cite	Total
Office of Naval Research (ONR)	314.0	168.2	482.2
Naval Sea Systems Command (NAVSEA)	19.6	19.7	39.3
Space and Naval Warfare Systems Command (SPAWAR)	34.6	8.1	42.6
Naval Air Systems Command (NAVAIR)	11.5	21.8	33.3
Other Navy	19.3	21.5	40.8
All Other	284.6	170.2	454.8
Total Funds	683.6	409.5	1,093.1

FY 2011
Source of Funds (%)

FY 2011	Reimbursable	$M Direct Cite	Total
Office of Naval Research (ONR)	319.6	101.6	421.2
Naval Sea Systems Command (NAVSEA)	22.9	10.9	33.8
Space and Naval Warfare Systems Command (SPAWAR)	40.7	54.7	95.4
Naval Air Systems Command (NAVAIR)	7.8	7.4	15.2
Other Navy	72.3	50.2	122.5
All Other	264.2	145.9	410.1
Total Funds	727.6	370.6	1,098.2

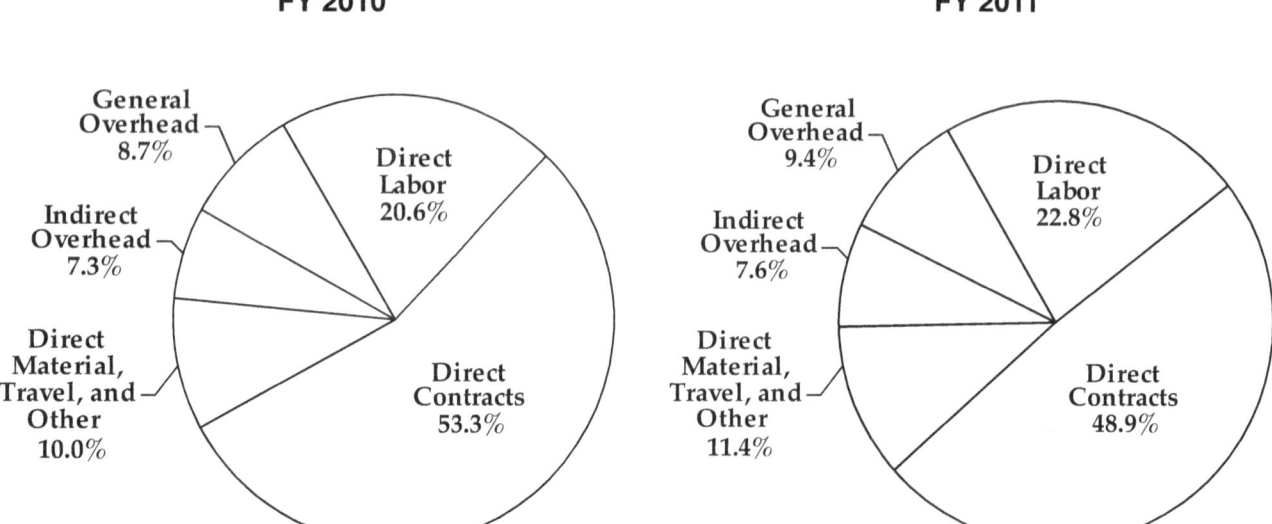

FY 2010

FY 2011

FY 2010

Distribution of Funds (%)

	$M
Direct Labor	228.9
General Overhead	96.0
Indirect Overhead	81.5
Direct Material, Travel, and Other	111.4
Direct Contracts	591.8
Total Costs*	1,109.6

FY 2011

Distribution of Funds (%)

	$M
Direct Labor	241.5
General Overhead	99.2
Indirect Overhead	80.0
Direct Material, Travel, and Other	120.3
Direct Contracts	517.3
Total Costs*	1,058.3

*Costs based on CFO statements; direct contracts include costs for reimbursable-funded contracts and obligations for direct cite-funded contracts.

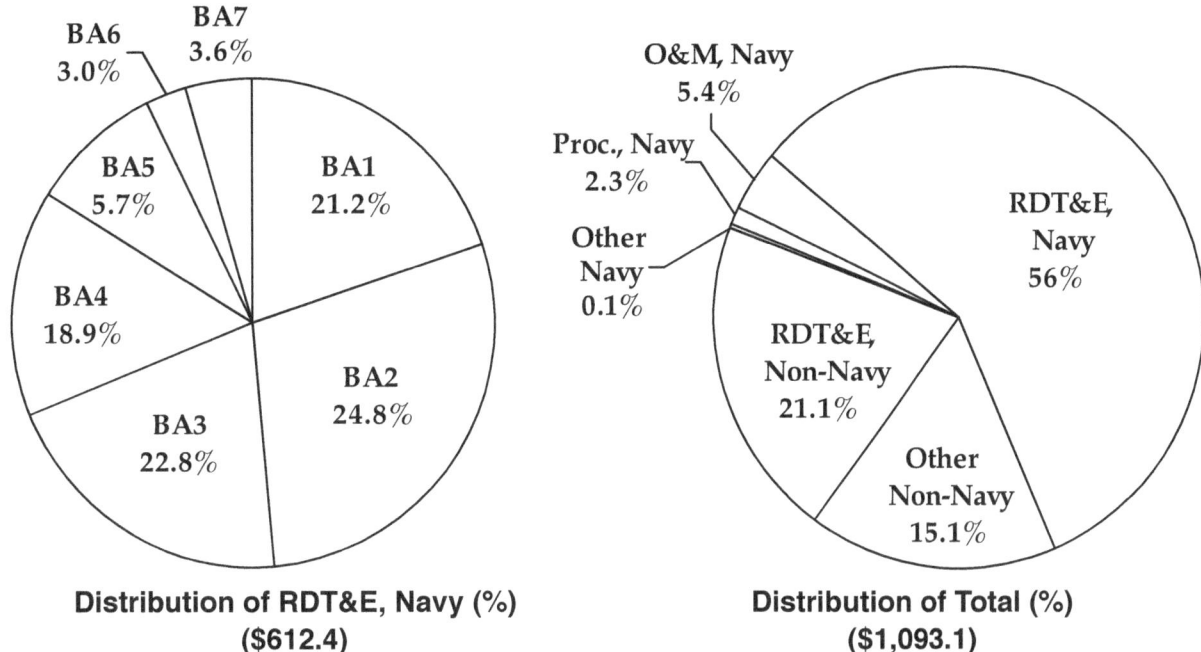

Distribution of RDT&E, Navy (%)
($612.4)

Distribution of Total (%)
($1,093.1)

FY 2010

Category	Navy	$M Non-Navy	Total
BA1 Basic Research	130.1	9.4	139.5
BA2 Applied Research	151.6	31.6	183.2
BA3 Advanced Technology Development	139.8	93.3	233.1
BA4 Advanced Component Development Prototypes	115.9	60.8	176.7
BA5 System Development and Demonstration	34.7	4.7	39.4
BA6 RDT&E Management Support	18.2	8.8	27
BA7 Operational System Development	22.1	22.4	44.5
Subtotal RDT&E	612.4	231	843.4
Operations and Maintenance	58.7	43.8	102.5
Procurement	24.7	39.5	64.2
Other	1.6	81.4	83
Total New Funds	697.4	395.7	1,093.1

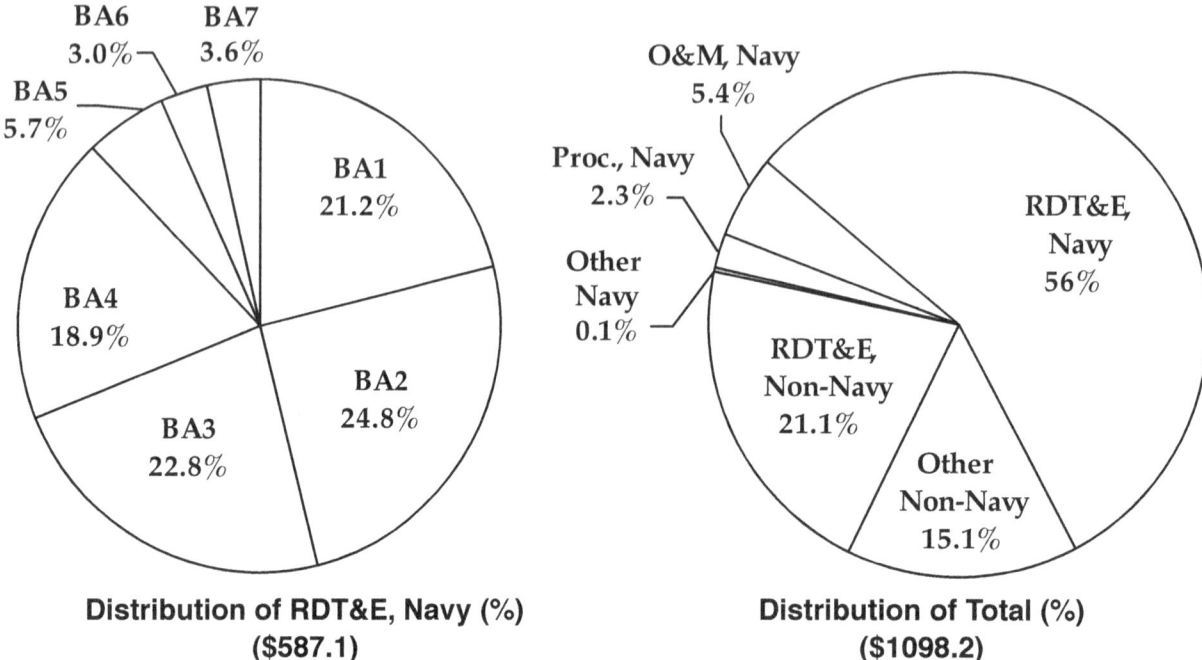

Distribution of RDT&E, Navy (%)
($587.1)

Distribution of Total (%)
($1098.2)

FY 2011

Category	Navy	$M Non-Navy	Total
BA1 Basic Research	126.7	7.4	134.1
BA2 Applied Research	161.2	33.5	194.7
BA3 Advanced Technology Development	117.3	90.4	207.7
BA4 Advanced Component Development Prototypes	112.1	67.6	179.7
BA5 System Development and Demonstration	31.6	15.5	47.1
BA6 RDT&E Management Support	17.2	7.2	24.4
BA7 Operational System Development	21	24.9	45.9
Subtotal RDT&E	587.1	246.5	833.6
Operations and Maintenance	65.7	58.5	102.5
Procurement	21.2	38.8	64.2
Other	1.7	78.7	83
Total New Funds	675.7	422.5	1,098.2

Personnel Information*

Civilian On-Board

Full-Time, Permanent (FTP)	
Graded	2,328
Ungraded	92
Total	2,420
Temporary, Part-Time, Intermittent (TPTI)	
TPTI	302
Total Civilian	2,722

FTP Breakdown

Scientific/Engineering Professional	1,572
Scientific/Engineering Technical	92
Administrative Specialist/Professional	376
Administrative Support	251
Senior Executive Service	21
Scientific or Professional	16
General Schedule	0
Total	2,328

Military On-Board

Officers	33
Enlisted	52
Total Military On-Board	85
(Military Allowance)	110

Annual Civilian Turnover Rate (%) (permanent employees only)

	2003	2004	2005	2006	2007	2008	2009	2010	2011
Research divisions	6.0	6.8	7.2	9.5	8.5	6.9	4.7	5	5.3
Nonresearch areas	8.2	8.2	8.5	11.0	13.7	13.3	7.4	11	13.5
Entire Laboratory	6.4	6.5	7.4	9.7	9.6	8.2	5.3	6.2	6.9

Highest Academic Degrees Held by Civilian Permanent Employees

Bachelors	573
Masters	368
Doctorates	839

*All data is as of 31 July 2012 unless otherwise noted.

**Professional
Development**

Programs for NRL Employees

The Human Resources Office supports and provides traditional and alternative methods of training for employees. NRL employees are encouraged to develop their skills and enhance their job performance so they can meet the future needs of NRL and achieve their own goals for growth.

One common study procedure is for employees to work full time at the Laboratory while taking job-related courses at universities and schools local to their job site. The training ranges from a single course to undergraduate, graduate, and postgraduate course work. Tuition for training is paid by NRL. The formal programs offered by NRL are described here.

Graduate Programs

The **Advanced Graduate Research Program** (formerly the Sabbatical Study Program, which began in 1964) enables selected professional employees to devote full time to research or pursue work in their own or a related field for up to one year at an institution or research facility of their choice without the loss of regular salary, leave, or fringe benefits. NRL pays all travel and moving expenses for the employee. Criteria for eligibility include professional stature consistent with the applicant's opportunities and experience, a satisfactory program of study, and acceptance by the facility selected by the applicant. The program is open to employees who have completed six years of Federal service, four of which have been at NRL.

The **Edison Memorial Graduate Training Program** enables employees to pursue graduate studies in their fields at local universities. Participants in this program may work 24 hours each workweek and pursue their studies during the other 16 hours. The criteria for eligibility include a minimum of one year of service at NRL, a bachelor's or master's degree in an appropriate field, and professional standing in keeping with the candidate's opportunities and experience.

To be eligible for the **Select Graduate Training Program**, employees must have a bachelor's degree in an appropriate field and must have demonstrated ability and aptitude for advanced training. Students accepted into this program receive one-half of their salary and benefits and NRL pays for tuition and travel expenses.

The **Naval Postgraduate School (NPS)**, located in Monterey, California, provides graduate programs to enhance the technical preparation of Naval officers and civilian employees who serve the Navy in the fields of science, engineering, operations analysis, and management. NRL employees desiring to pursue graduate studies at NPS may apply; thesis work is accomplished at NRL. Participants continue to receive full pay and benefits during the period of study. NRL also pays for tuition and travel expenses.

In addition to NRL and university offerings, application may be made to a number of noteworthy programs and fellowships. Examples of such opportunities are the **Capitol Hill Workshops**, the **Legislative Fellowship (LEGIS) program**, the **Federal Executive Institute (FEI)**, and the **Executive Leadership Program for Mid-Level Employees**. These and other programs are announced from time to time, as schedules are published.

Continuing Education

Undergraduate and graduate courses offered at local colleges and universities may be subsidized by NRL for employees interested in improving their skills and keeping abreast of current developments in their fields.

NRL offers **short courses** to all employees in a number of fields of interest including administrative subjects and supervisory and management techniques. Laboratory employees may also attend these courses at nongovernment facilities.

For further information on any of the above Graduate and Continuing Education programs, contact the Workforce Development and Management Branch (Code 1840) at (202) 404-8314 or via email at Training@hro.nrl.navy.mil.

The **Scientist-to-Sea Program (STSP)** provides opportunities for Navy R&D laboratory/center personnel to go to sea to gain first-hand insight into operational factors affecting system design, performance, and operations on a variety of ships. NRL is a participant of this Office of Naval Research (ONR) program. Contact (202) 767-7627.

Professional Development

NRL has several programs, professional society chapters, and informal clubs that enhance the professional growth of employees. Some of these are listed below.

The **Counseling & Referral Service (C/RS)** helps employees improve job performance through counseling designed to resolve problems that may adversely affect job performance. Such problems may include family and/or work-related stress, relationship difficulties, or behavioral, emotional, or substance abuse problems. C/RS provides confidential assessment, short-term counseling, training workshops, and referral to additional resources in the community. Contact (202) 767-6857.

The NRL **Women in Science and Engineering (WISE) Network** was formed in 1997 through the merger of the NRL chapter of WISE and the Women in Science and Technology Network. Luncheon meetings and seminars are held to discuss scientific research areas, career opportunities, and career-building strategies. The group also sponsors projects to promote the professional success of the NRL S&T community and improve the NRL working environment. Membership is open to all S&T professionals. Contact (202) 404-4389.

Sigma Xi, The Scientific Research Society, encourages and acknowledges original investigation in pure and applied science. It is an honor society for research scientists. Individuals who have demonstrated the ability to perform original research are elected to membership in local chapters. The NRL Edison Chapter, comprising approximately 200 members, recognizes original research by presenting annual awards in pure and applied science to two outstanding NRL staff members per year. In addition, an award seeking to reward rising stars in the lab is presented annuallly through the Young Investigator Award. The chapter also sponsors several lectures per year at NRL on a wide range of topics of general interest to the scientific and DoD community. These lectures are delivered by scientists from all over the world. The highlight of the Sigma Xi Lecture Series is the Edison Memorial Lecture, which traditionally is given by a internationally distinguished scientist. Contact (202) 767-2007.

The **NRL Mentor Program** was established to provide an innovative approach to professional and career training and an environment for personal and professional growth. It is open to permanent NRL employees in all job series and at all sites. Mentorees are matched with successful, experienced colleagues having more technical and/or managerial experience who can provide them with the knowledge and skills needed to maximize their contribution to the success of their immediate organization, to NRL, to the Navy, and to their chosen career fields. The ultimate goal of the program is to increase job productivity, creativity, and satisfaction through better communication, understanding, and training. NRL Instruction 12400.1B provides policy and procedures for the program. For more information please contact mentor@hro.nrl.navy.mil or (202) 767-6736.

Employees interested in developing effective self-expression, listening, thinking, and leadership potential are invited to join the Forum Club, a chapter of **Toastmasters International**. Members of this club possess diverse career backgrounds and talents and learn to communicate not by rules but by practice in an atmosphere of understanding and helpful fellowship. NRL's Commanding Officer and Director of Research endorse Toastmasters. Contact (202) 404-4670.

Equal Employment Opportunity (EEO) Programs

Equal employment opportunity (EEO) is a fundamental NRL policy for all employees regardless of race, color, national origin, sex, religion, age, sexual orientation, or disability. The NRL EEO Office is a service organization whose major functions include counseling employees in an effort to resolve employee/management conflicts, processing formal discrimination complaints, providing EEO training, and managing NRL's affirmative employment recruitment program. The NRL EEO Office is also responsible for sponsoring special-emphasis programs to promote awareness and increase sensitivity and appreciation of the issues or the history relating to females, individuals with disabilities, and minorities. Contact the NRL Deputy EEO Officer at (202) 767-2486 for additional information on any of their programs or services.

Other Activities

The award-winning **Community Outreach Program** directed by the NRL Public Affairs Office fosters programs that benefit students and other community citizens. Volunteer employees assist with and judge science fairs, give lectures, provide science demonstrations and student tours of NRL, and serve as tutors, mentors, coaches, and classroom resource teachers. The program sponsors student tours of NRL, and an annual holiday party for neighborhood children in December. Through the program, NRL has active partnerships with three District of Columbia public schools. Contact (202) 767-2541.

Other programs that enhance the development of NRL employees include sports and theater groups and the **Amateur Radio Club**. The **NRL Fitness Center** at NRL-DC, managed by Naval Support Activity Washington Morale, Welfare and Recreation (NSAW-MWR), houses a fitness room with treadmills, bikes, ellipticals, step mills, and a full strength circuit; a gymnasium for basketball, volleyball, and other activities; a game room; and full locker rooms. The Fitness Center is free to NRL employees and contractors. NRL employees are also eligible to participate in all NSAW MWR activities held on Joint Base Anacostia–Bolling and Washington Navy Yard, less than five miles away. The **NRL Showboaters Theatre**, organized in 1974, is "in the dark." Visit www.nrl.navy. mil/ showboaters/Past_Productions.php for pictures from past productions such as Annie Get Your Gun, Gigi, and Hello Dolly. Contact (202) 404-4998 for Play Reader's meetings at NRL.

Programs for Non-NRL Employees

Several programs have been established for non-NRL professionals. These programs encourage and support the participation of visiting scientists and engineers in research of interest to the Laboratory. Some of the programs may serve as stepping-stones to Federal careers in science and technology. Their objective is to enhance the quality of the Laboratory's research activities through working associations and interchanges with highly capable scientists and engineers and to provide opportunities for outside scientists and engineers to work in the Navy laboratory environment. Along with enhancing the Laboratory's research, these programs acquaint participants with Navy capabilities and concerns and may provide a path to full-time employment.

Recent Ph.D., Faculty Member, and College Graduate Programs

The **National Research Council (NRC) Cooperative Research Associateship Program** selects associates who conduct research at NRL in their chosen fields in collaboration with NRL scientists and engineers. Appointments are for one year (renewable for a second and possible third year).

The **NRL/ASEE Postdoctoral Fellowship Program,** administered by the American Society for Engineering Education (ASEE), aims to increase the involvement of highly trained scientists and engineers in disciplines necessary to meet the evolving needs of naval technology. Appointments are for one year (renewable for a second and possible third year).

The **Naval Research Enterprise Intern Program (NREIP)** is a ten-week program involving NROTC colleges/universities and their affiliates. The Office of Naval Research (ONR) offers summer appointments at Navy laboratories to current sophomores, juniors, seniors, and graduate students from participating schools. Application is online at www.asee.org/nreip through the American Society for Engineering Education. Electronic applications are sent for evaluation to the point of contact at the Navy laboratory identified by the applicant. Students are provided a stipend of $7,500 (undergraduates) or $10,000 (graduate students).

The American Society for Engineering Education also administers the **Navy/ASEE Summer Faculty Research and Sabbatical Leave Program** for university faculty members to work for ten weeks (or longer, for those eligible for sabbatical leave) with professional peers in participating Navy laboratories on research of mutual interest.

The **NRL/United States Naval Academy (USNA) Cooperative Program for Scientific Interchange** allows faculty members of the U.S. Naval Academy to participate in NRL research. This collaboration benefits the Academy by providing the opportunity for USNA faculty members to work on research of a more practical or applied nature. In turn, NRL's research program is strengthened by the available scientific and engineering expertise of the USNA faculty.

The **National Defense Science and Engineering Graduate Fellowship Program** helps U.S. citizens obtain advanced training in disciplines of science and engineering critical to the U.S. Navy. The three-year program awards fellowships to recent outstanding graduates to support their study and research leading to doctoral degrees in specified disciplines such as electrical engineering, computer sciences, material sciences, applied physics, and ocean engineering. Award recipients are encouraged to continue their study and research in a Navy laboratory during the summer.

For further information about the above six programs, contact (202) 404-7450.

Professional Appointments

Faculty Member Appointments use the special skills and abilities of faculty members for short periods to fill positions of a scientific, engineering, professional, or analytical nature at NRL.

Consultants and experts are employed because they are outstanding in their fields of specialization or because they possess ability of a rare nature and could not normally be employed as regular civil servants.

Intergovernmental Personnel Act Appointments temporarily assign personnel from state or local governments or educational institutions to the Federal Government (or vice versa) to improve public services rendered by all levels of government.

College and High School Student Programs

The student programs are tailored to high school, undergraduate, and graduate students to provide employment opportunities and work experience in naval research. These programs are designed to attract applicants for student and full professional employment in fields such as engineering, physics, mathematics, and computer sciences. The student employment programs are designed to help students and educational institutions gain a better understanding of NRL's research, its challenges, and its opportunities. To participate in these programs, the student must be continuously enrolled in school on at least a half-time basis at a qualifying educational institution; and be at least 16 years of age and a U.S. citizen.

The **Student Career Experience Program (SCEP)** employs students in study-related occupations. The program is conducted in accordance with a planned schedule and a working agreement among NRL, the educational institution, and the student. Primary focus is on the pursuit of undergraduate and graduate degrees in engineering, computer science, or the physical sciences. Applications are accepted year-round.

The **Student Temporary Employment Program (STEP)** is a one year temporary employment program that may be renewed. This program enables students to earn a salary while continuing their studies and offers them valuable work experience. They must be continuously enrolled in school on at least a half-time basis at a qualifying educational institution. Applications are accepted year-round.

The **Summer Employment Program (SEP)** employs students for the summer that are enrolled in a qualifying educational institution on at least a half-time basis studying paraprofessional and technician positions in engineering, physical sciences, computer sciences, and mathematics. Applications are due the second Friday in February.

The **Student Volunteer Program** helps students gain valuable experience by allowing them to voluntarily perform educationally related work at NRL. Applications are accepted year-round.

For additional information on these student programs, contact (202) 767-8313.

For high school students, the **DoD Science & Engineering Apprentice Program (SEAP)** offers students grades 9 to 12 the opportunity to serve for eight weeks to participate in research at a Department of Navy laboratory during the summer. Under the direction of a mentor, students gain a better understanding of the challenges and opportunities of research through participation in scientific programs. Criteria for eligibility are based on science and mathematics courses completed and grades achieved; scientific motivation, curiosity, and capacity for sustained hard work; a desire for a technical career; teacher recommendations; and achievement test scores. For more information, please contact the SEAP coordinator at SEAP@hro.nrl.navy.mil or (202) 767-8324/8309/6736.

General Information

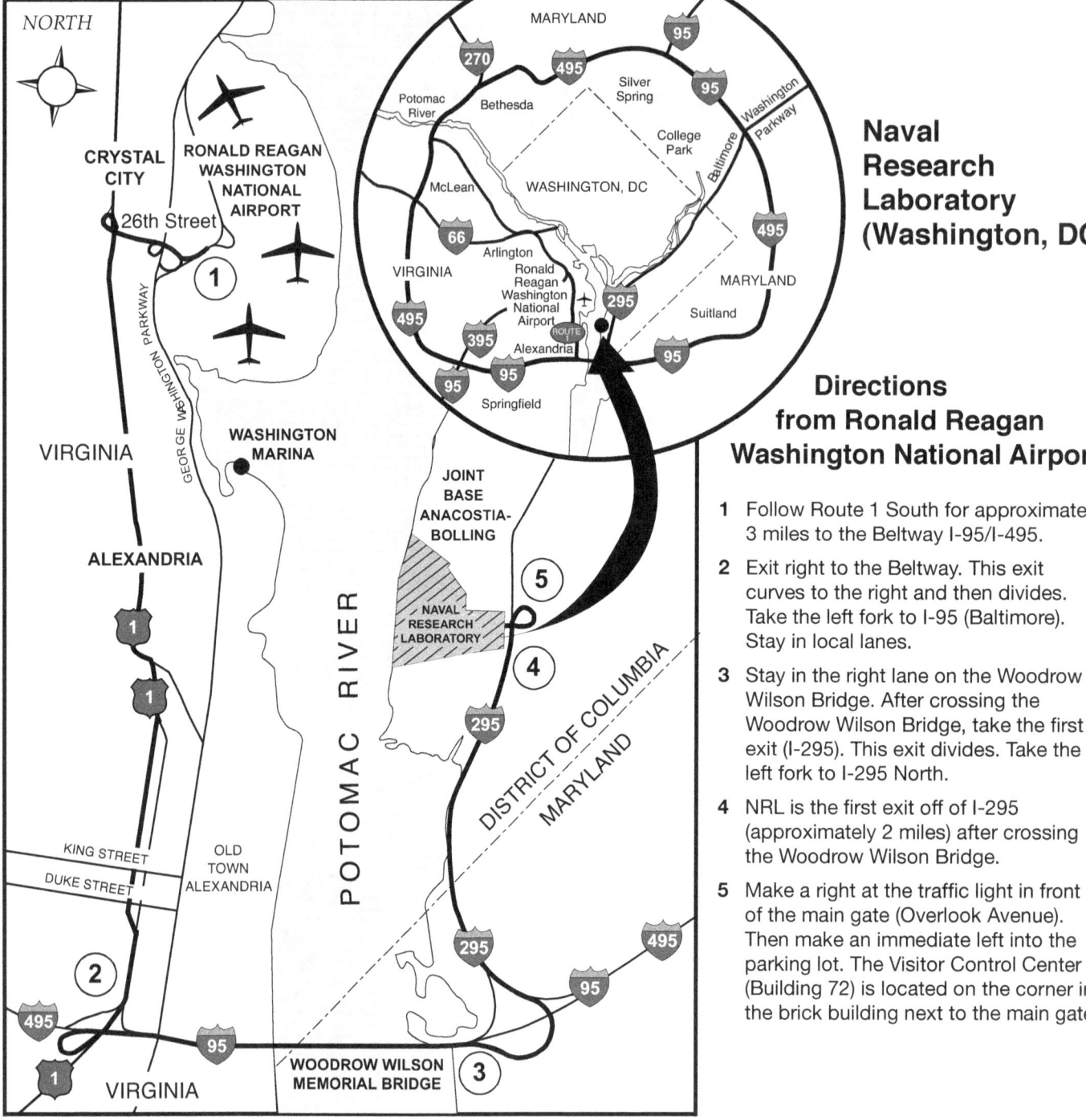

Naval Research Laboratory (Washington, DC)

Directions from Ronald Reagan Washington National Airport

1 Follow Route 1 South for approximately 3 miles to the Beltway I-95/I-495.

2 Exit right to the Beltway. This exit curves to the right and then divides. Take the left fork to I-95 (Baltimore). Stay in local lanes.

3 Stay in the right lane on the Woodrow Wilson Bridge. After crossing the Woodrow Wilson Bridge, take the first exit (I-295). This exit divides. Take the left fork to I-295 North.

4 NRL is the first exit off of I-295 (approximately 2 miles) after crossing the Woodrow Wilson Bridge.

5 Make a right at the traffic light in front of the main gate (Overlook Avenue). Then make an immediate left into the parking lot. The Visitor Control Center (Building 72) is located on the corner in the brick building next to the main gate.

Naval Research Laboratory
4555 Overlook Avenue, SW
Washington, DC 20375-5320
(202) 767-3200 – DSN 297-3200

Location of Buildings at NRL Washington

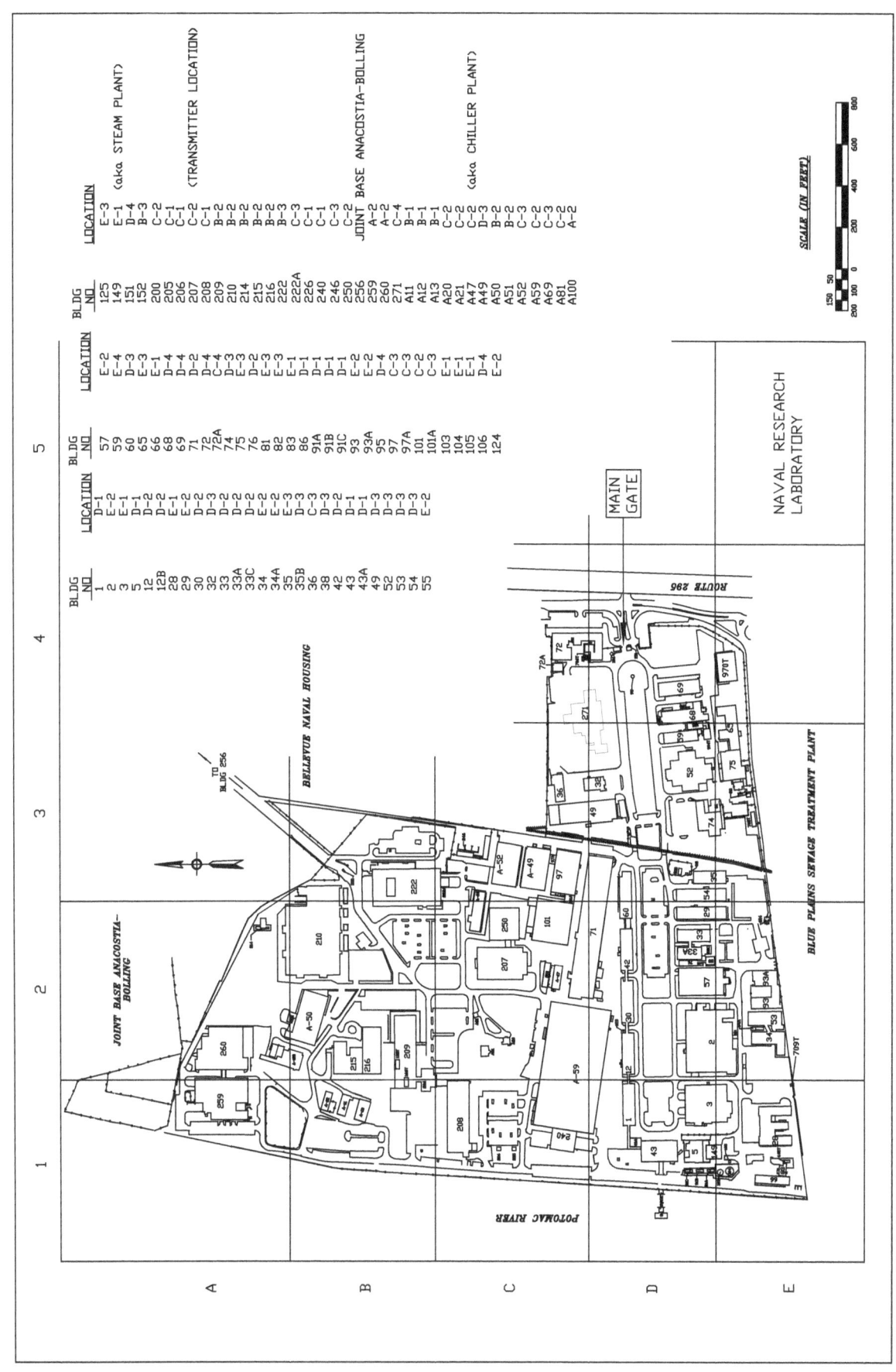

Location of Field Sites in the NRL Washington Area

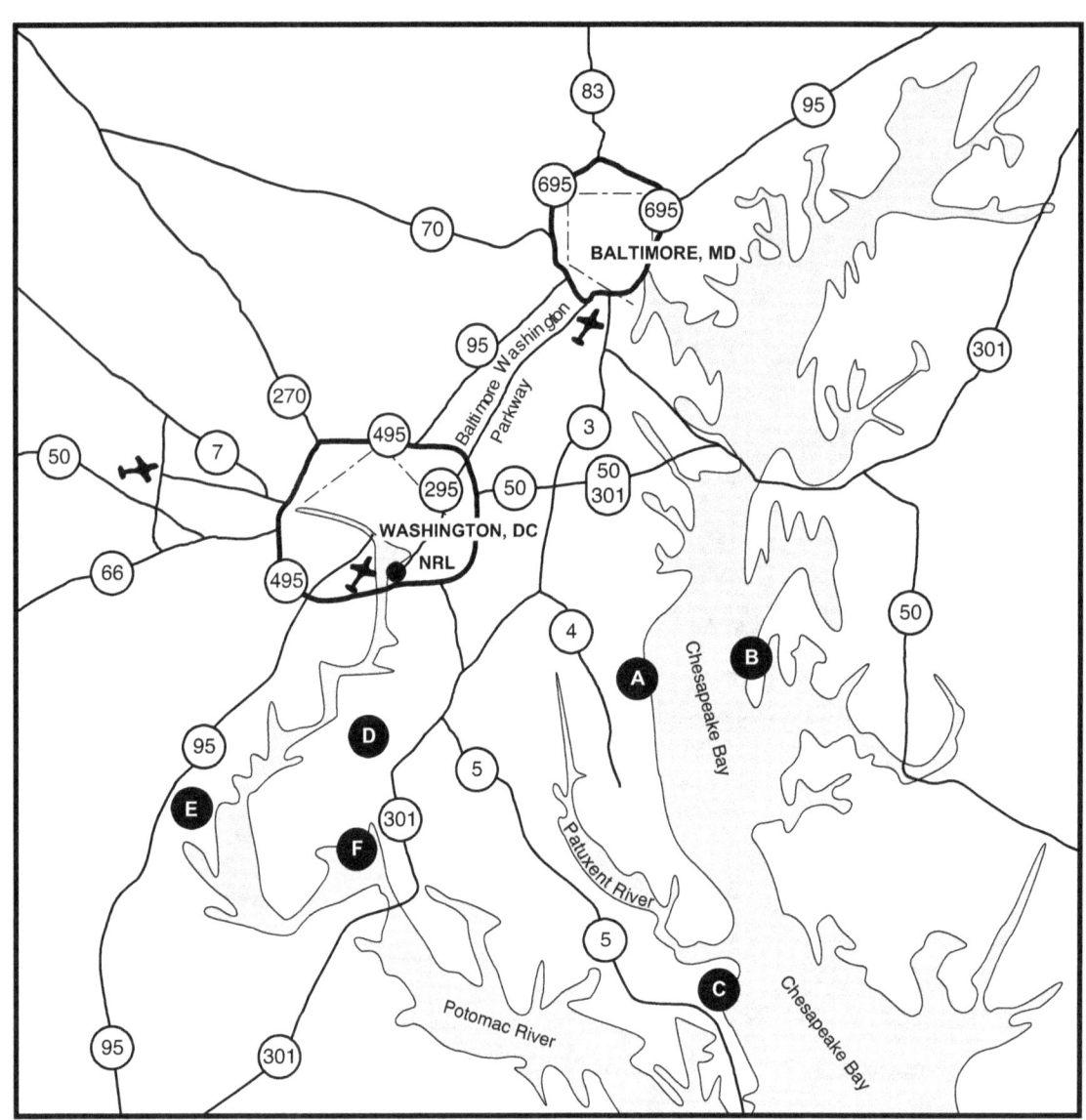

	Location	Approximate Mileage from NRL Washington	Cognizant Code
A –	Chesapeake Bay Section, Chesapeake Beach, MD	40	3522
B –	Tilghman Island, MD	110	3522
C –	Patuxent River (MD) Naval Air Station	64	1600
D –	Pomonkey, MD	20	8124
E –	Midway Research Center, Quantico, VA	38	8140
F –	Blossom Point, MD	40	8140

Chesapeake Bay Section
(Chesapeake Beach, Maryland)

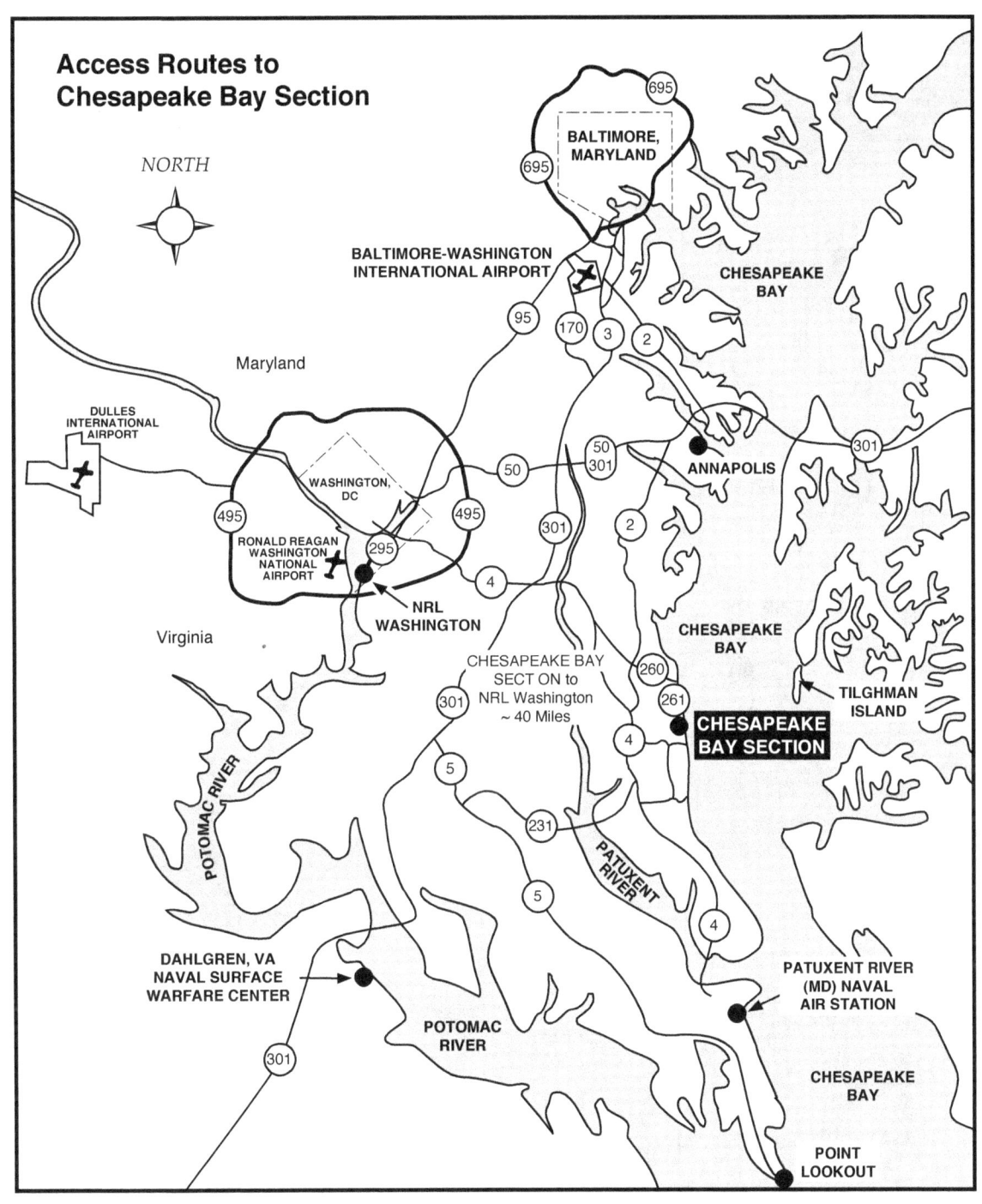

Naval Research Laboratory
Chesapeake Bay Section
5813 Bayside Road
Chesapeake Beach, MD 20732
(301) 257-4002

Location of Buildings
at the Chesapeake Bay Section

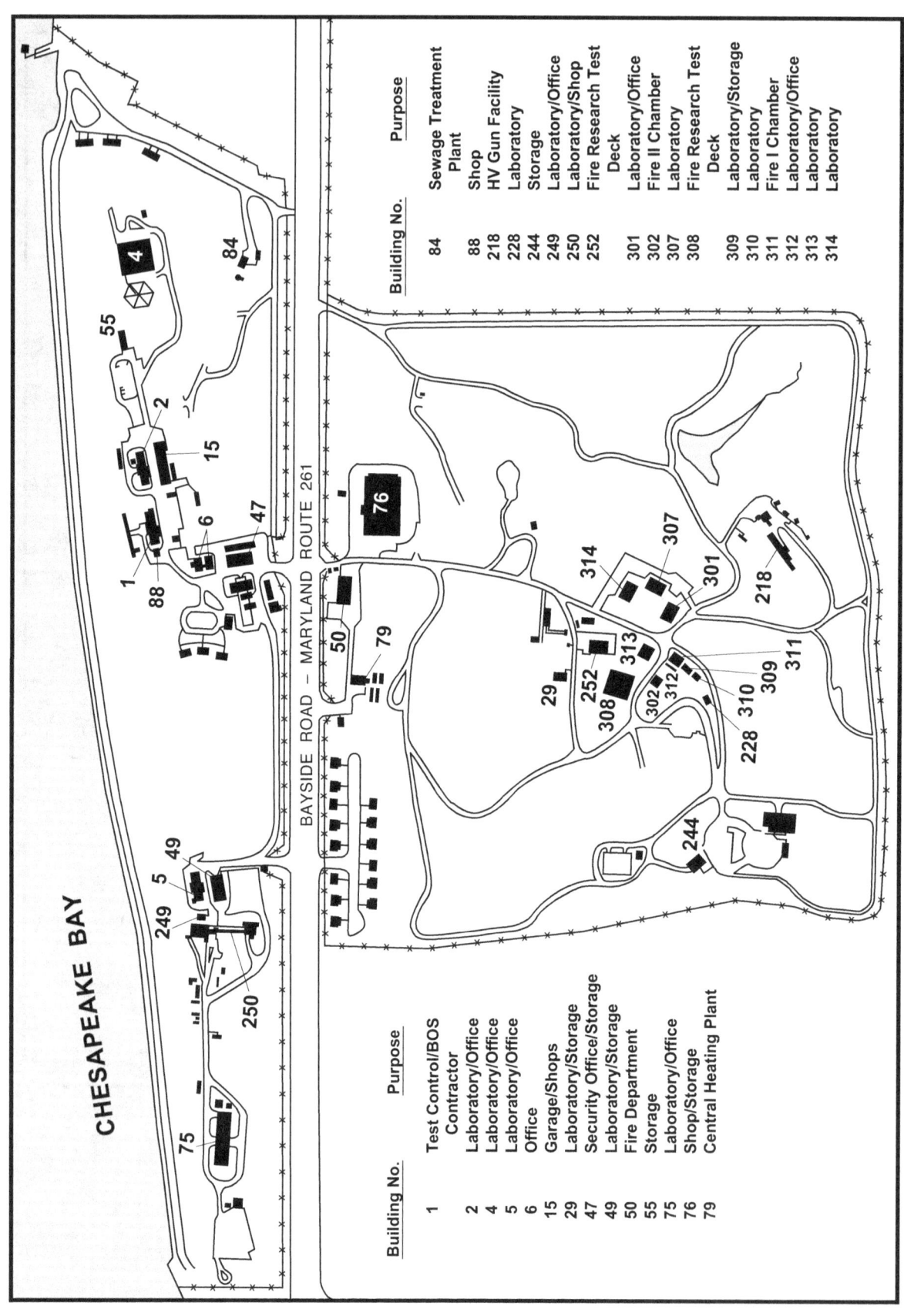

Building No.	Purpose
84	Sewage Treatment Plant
88	Shop
218	HV Gun Facility
228	Laboratory
244	Storage
249	Laboratory/Office
250	Laboratory/Shop
252	Fire Research Test Deck
301	Laboratory/Office
302	Fire II Chamber
307	Laboratory
308	Fire Research Test Deck
309	Laboratory/Storage
310	Laboratory
311	Fire I Chamber
312	Laboratory/Office
313	Laboratory
314	Laboratory

Building No.	Purpose
1	Test Control/BOS Contractor
2	Laboratory/Office
4	Laboratory/Office
5	Laboratory/Office
6	Office
15	Garage/Shops
29	Laboratory/Storage
47	Security Office/Storage
49	Laboratory/Storage
50	Fire Department
55	Storage
75	Laboratory/Office
76	Shop/Storage
79	Central Heating Plant

BAYSIDE ROAD – MARYLAND ROUTE 261

CHESAPEAKE BAY

John C. Stennis Space Center
(Stennis Space Center, Mississippi)

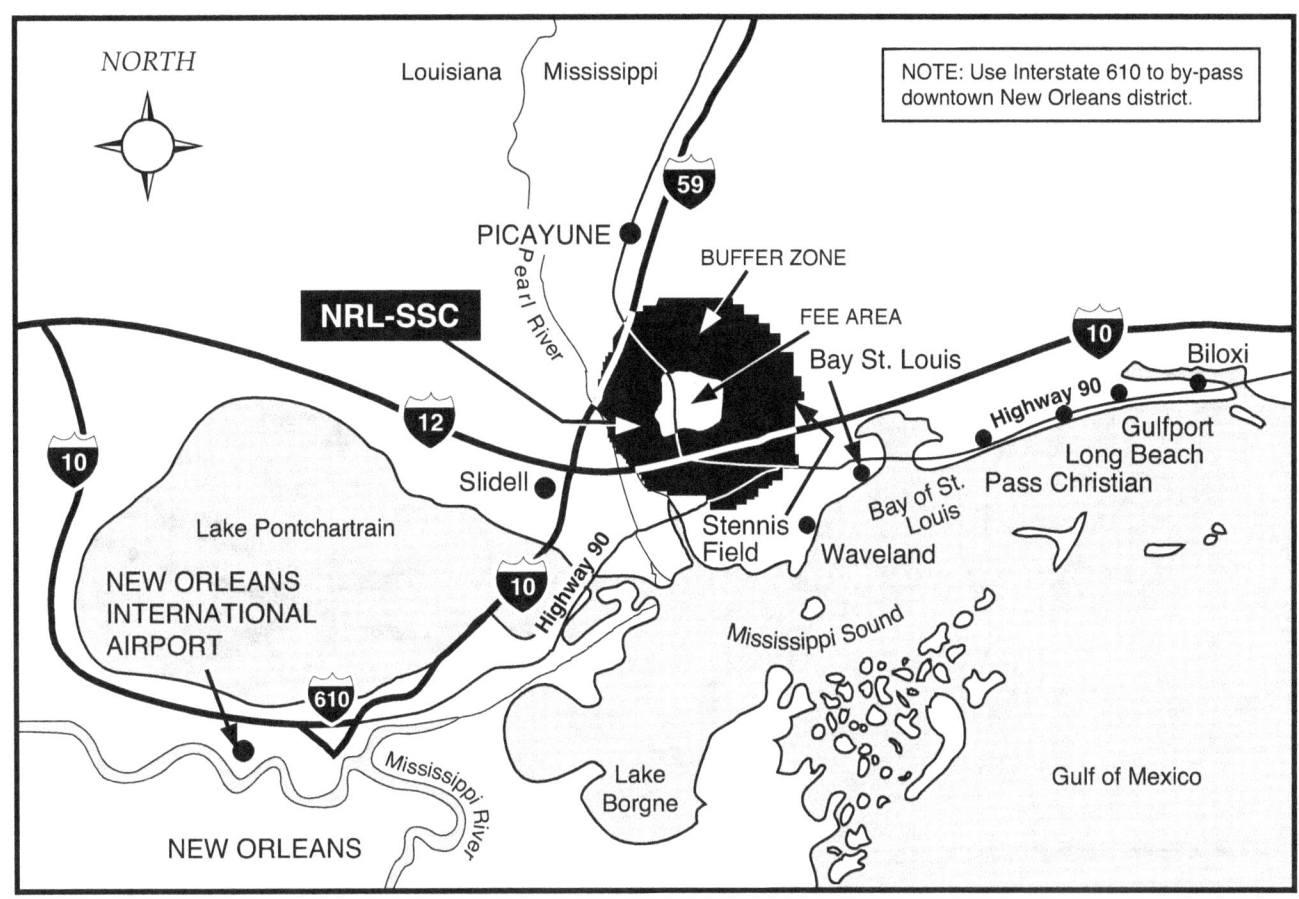

Naval Research Laboratory
John C. Stennis Space Center
Stennis Space Center, MS 39529-5004
(228) 688-3390

Naval Research Laboratory Monterey
(Monterey, California)

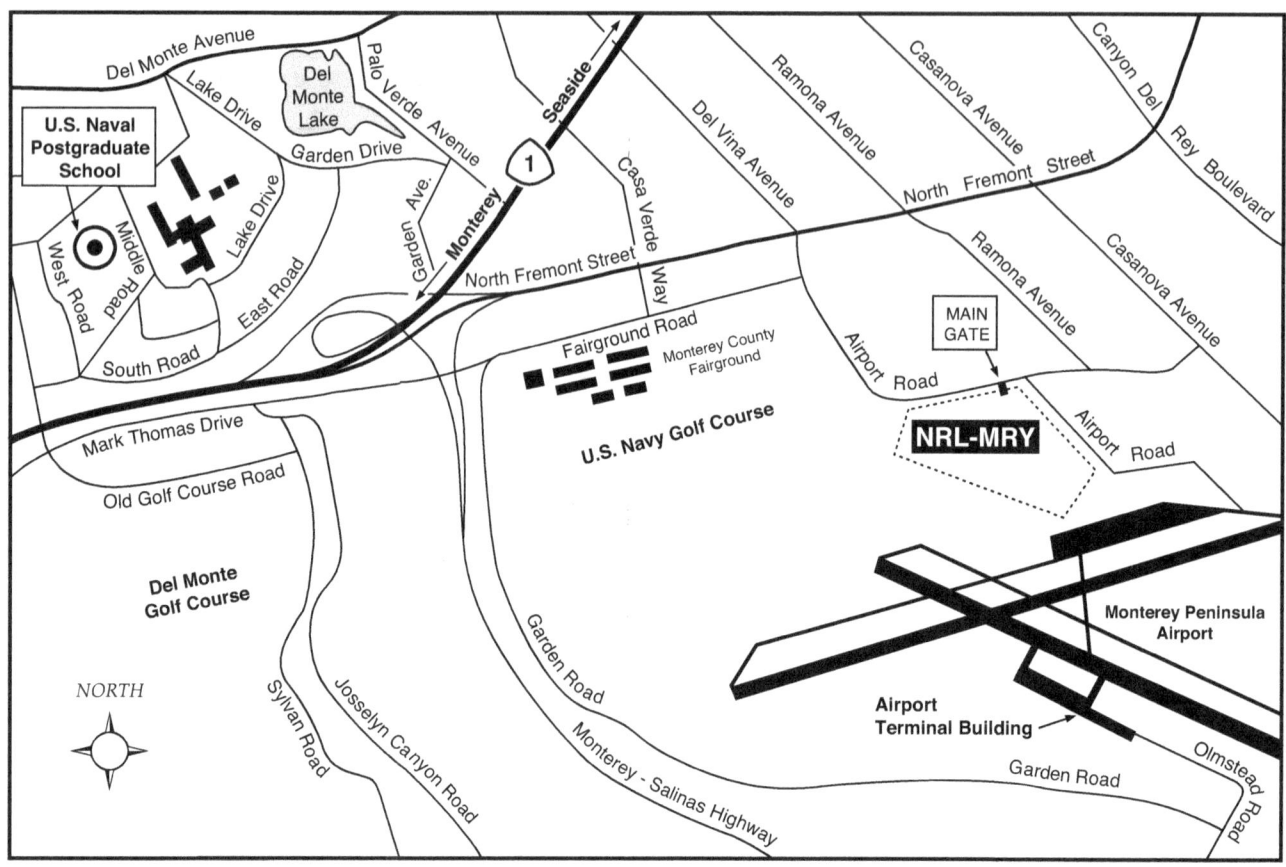

Naval Research Laboratory
Marine Meteorology Division
7 Grace Hopper Avenue
Monterey, CA 93943-5502
(831) 656-4721

Key Personnel

Code		Telephone

EXECUTIVE DIRECTORATE

Code		Telephone
1000	Commanding Officer	(202) 767-3403
1000.1	Inspector General	(202) 767-3621
1001	Director of Research	(202) 767-3301
1001.1	Executive Assistant to the Director of Research	(202) 767-2445
1001.2	Head, Strategic Workforce Planning	(202) 767-3421
1001.3	Executive Assistant for Technology Deployment	(202) 767-0851
1002	Chief Staff Officer	(202) 767-3621
1004	Head, Office of Technology Transfer	(202) 767-3083
1006	Head, Office of Program Administration and Policy Development	(202) 767-1312
1008	Head, Office of Counsel	(202) 767-2244
1030	Head, Public Affairs Office	(202) 767-2541
1100	Director, Institute for Nanoscience	(202) 767-1803
1200	Head, Command Support Division	(202) 767-3091
1400	Head, Military Support Division	(202) 767-2273
1600	Commanding Officer, Scientific Development Squadron One (PAX River NAS)	(301) 342-3751
1700	Director, Laboratory for Autonomous Systems Research	(202) 767-0792
1800	Director, Human Resources Office	(202) 767-8322
1830	Deputy Equal Employment Opportunity Officer	(202) 767-8390
3005	Deputy for Small Business	(202) 767-0666
3540	Head, Safety Branch	(202) 767-2232

BUSINESS OPERATIONS DIRECTORATE

Code		Telephone
3000	Associate Director of Research for Business Operations	(202) 767-2371
3005	Deputy for Small Business	(202) 767-0666
3030	Head, Management Information Systems Office	(202) 404-3659
3200	Head, Contracting Division	(202) 767-5227
3300	Head, Financial Management Division	(202) 767-3405
3400	Head, Supply and Information Services Division	(202) 767-3446
3500	Director, Research and Development Services Division	(202) 404-4054

SYSTEMS DIRECTORATE

Code		Telephone
5000	Associate Director of Research for Systems	(202) 767-3525
5300	Superintendent, Radar Division	(202) 404-2700
5500	Superintendent, Information Technology Division/NRL Chief Information Officer*	(202) 767-2903
5600	Superintendent, Optical Sciences Division	(202) 767-3171
5700	Superintendent, Tactical Electronic Warfare Division	(202) 767-6278

MATERIALS SCIENCE AND COMPONENT TECHNOLOGY DIRECTORATE

Code		Telephone
6000	Associate Director of Research for Materials Science and Component Technology	(202) 767-3566
6100	Superintendent, Chemistry Division	(202) 767-3026
6300	Superintendent, Materials Science and Technology Division	(202) 767-2926
6040	Director, Laboratories for Computational Physics and Fluid Dynamics	(202) 767-3055
6700	Superintendent, Plasma Physics Division	(202) 767-2723
6800	Superintendent, Electronics Science and Technology Division	(202) 767-3693
6900	Director, Center for Bio/Molecular Science and Engineering	(202) 404-6000

---

*Additional duty

Code   **Telephone**

OCEAN AND ATMOSPHERIC SCIENCE AND TECHNOLOGY DIRECTORATE

Code		Telephone
7000	Associate Director of Research for Ocean and Atmospheric Science and Technology	(202) 404-8690
7030	Head, Office of Research Support Services	(228) 688-4010
7100	Superintendent, Acoustics Division	(202) 767-3482
7200	Superintendent, Remote Sensing Division	(202) 767-3391
7300	Superintendent, Oceanography Division	(228) 688-4670
7400	Superintendent, Marine Geosciences Division	(228) 688-4650
7500	Superintendent, Marine Meteorology Division	(831) 656-4721
7600	Superintendent, Space Science Division	(202) 767-6343

NAVAL CENTER FOR SPACE TECHNOLOGY

Code		Telephone
8000	Director, Naval Center for Space Technology	(202) 767-6547
8100	Superintendent, Space Systems Development Department	(202) 767-4593
8200	Superintendent, Spacecraft Engineering Department	(202) 404-3727

www.ingramcontent.com/pod-product-compliance
Lightning Source LLC
Chambersburg PA
CBHW081203210526
45170CB00025B/2058

Code 5600
Staff Activity Areas

Research to develop a fundamental physical understanding of highly variable transient and long-term solar activity; the radiative, plasma, and particulate emissions associated with the activity; and the responses of the heliosphere and the terrestrial magnetosphere to the activity. Relevant empirical data is collected by conceiving, developing, and operating state-of-the-art imaging, spectrometric, and in situ space flight sensors on national and international space missions. Physics-based models are hypothesized, tested with the collected empirical data and computer simulation, and developed.

SECCHI: The Sun-Earth Connection and Heliospheric Investigation instrument suite, shown during testing at NRL, is returning spectacular stereo imagery of the region between the Sun and the Earth.

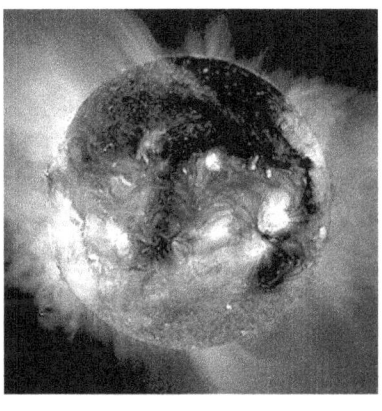

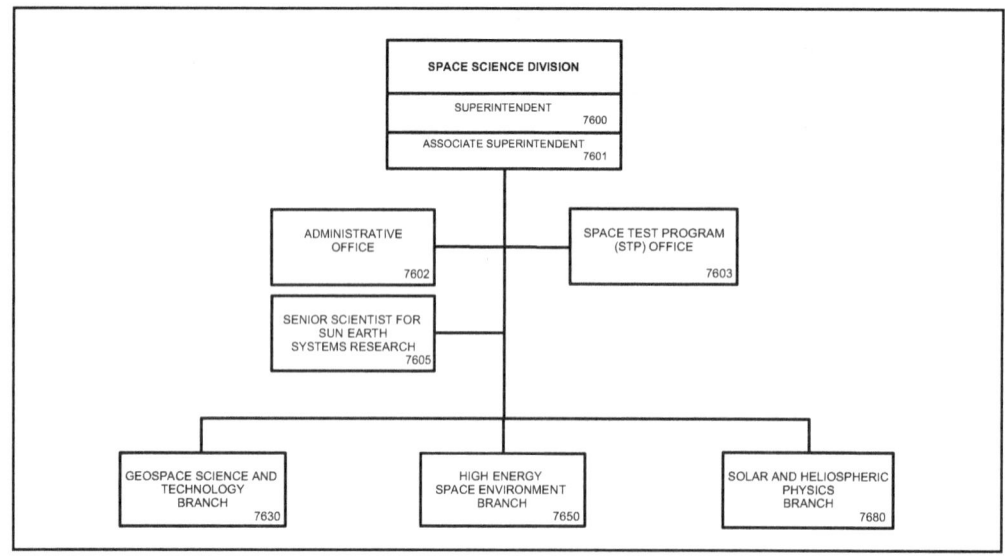

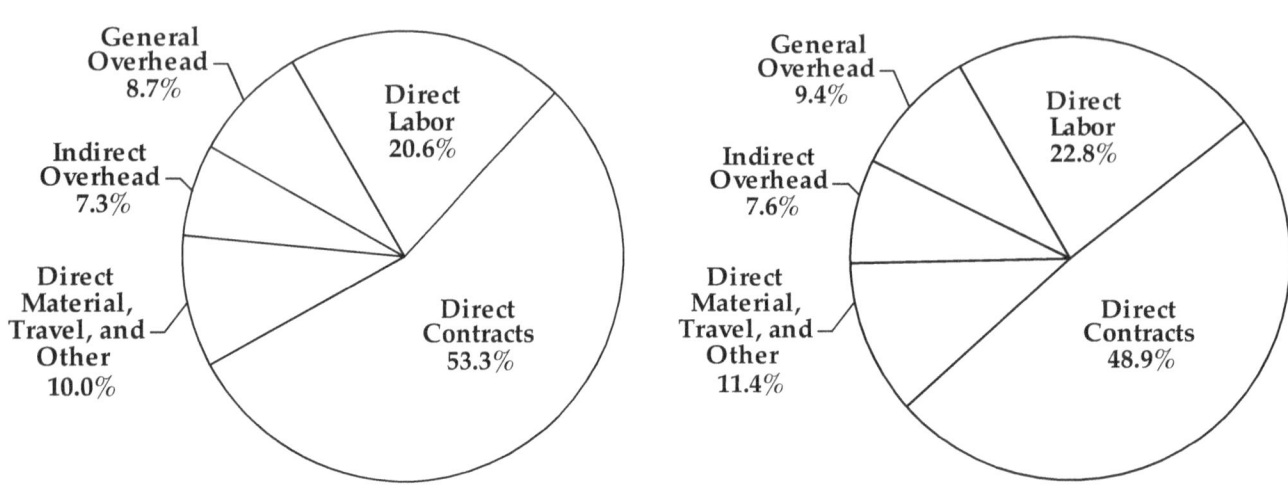

BLDG NO	LOCATION
1	D-1
2	E-2
3	E-1
5	D-1
12	D-2
12B	D-2
28	E-1
29	E-2
30	D-2
32	D-2
33	D-2
33A	D-2
33C	D-2
34	D-2
34A	E-2
35	E-3
35B	C-3
36	D-2
38	D-2
42	D-1
43	D-1
43A	D-1
49	D-3
52	D-3
53	D-3
54	D-3
55	E-2

BLDG NO	LOCATION
57	E-2
59	E-4
60	E-3
65	E-1
66	D-4
68	D-4
69	D-4
71	C-2
72A	D-4
74	D-4
75	D-3
76	D-2
81	D-2
82	E-3
83	E-1
86	D-1
91A	D-1
91B	D-1
91C	D-1
93	E-2
93A	E-2
95	D-4
97	C-3
97A	C-2
101	C-3
101A	E-1
103	E-1
104	E-1
105	D-4
106	D-4
124	E-2

BLDG NO	LOCATION	
125	E-3	(aka STEAM PLANT)
149	E-1	
151	D-4	
152	B-3	
200	C-2	
205	C-1	
206	C-1	(TRANSMITTER LOCATION)
207	C-1	
208	B-2	
209	B-2	
210	B-2	
214	B-2	
215	B-3	
216	B-3	
222	C-3	
222A	C-1	
226	C-1	
240	C-3	
246	C-2	
250	C-2	
256	B-2	JOINT BASE ANACOSTIA-BOLLING
259	A-2	
260	A-2	
271	C-4	
A11	B-1	
A12	B-1	
A13	B-1	
A20	C-2	
A21	C-2	
A47	C-2	
A49	D-3	
A50	A-2	
A51	B-2	
A52	C-3	
A59	C-3	
A69	C-3	
A81	C-3	(aka CHILLER PLANT)
A100	A-2	

BELLEVUE NAVAL HOUSING

JOINT BASE ANACOSTIA-BOLLING

BLUE PLAINS SEWAGE TREATMENT PLANT

DSN: NRL Washington 297- or 754-; NRL/SSC 828-; NRL/Monterey 878-;
NRL VXS-1/Patuxent River 342-